Probability and Mathematical Statistics (Continu

LEHMANN • Theory of Point Estimation
MATTHES, KERSTAN, and MECKE • In
MUIRHEAD • Aspects of Multivariate Sta
PRESS • Bayesian Statistics: Principles, Mo
PURI and SEN • Nonparametric Methods i
PURI and SEN • Nonparametric Methods in Multivariate Analysis
PURI, VILAPLANA, and WERTZ • New Perspectives in Theoretical and Applied Statistics
RANDLES and WOLFE • Introduction to the Theory of Nonparametric Statistics
RAO • Linear Statistical Inference and Its Applications, *Second Edition*
RAO • Real and Stochastic Analysis
RAO and SEDRANSK • W.G. Cochran's Impact on Statistics
RAO • Asymptotic Theory of Statistical Inference
ROBERTSON, WRIGHT and DYKSTRA • Order Restricted Statistical Inference
ROGERS and WILLIAMS • Diffusions, Markov Processes, and Martingales,Volume II: Îto Calculus
ROHATGI • An Introduction to Probability Theory and Mathematical Statistics
ROHATGI • Statistical Inference
ROSS • Stochastic Processes
RUBINSTEIN • Simulation and The Monte Carlo Method
RUZSA and SZEKELY • Algebraic Probability Theory
SCHEFFE • The Analysis of Variance
SEBER • Linear Regression Analysis
SEBER • Multivariate Observations
SEBER and WILD • Nonlinear Regression
SEN • Sequential Nonparametrics: Invariance Principles and Statistical Inference
SERFLING • Approximation Theorems of Mathematical Statistics
SHORACK and WELLNER • Empirical Processes with Applications to Statistics
STOYANOV • Counterexamples in Probability

Applied Probability and Statistics

ABRAHAM and LEDOLTER • Statistical Methods for Forecasting
AGRESTI • Analysis of Ordinal Categorical Data
AICKIN • Linear Statistical Analysis of Discrete Data
ANDERSON and LOYNES • The Teaching of Practical Statistics
ANDERSON, AUQUIER, HAUCK, OAKES, VANDAELE, and WEISBERG • Statistical Methods for Comparative Studies
ARTHANARI and DODGE • Mathematical Programming in Statistics
ASMUSSEN • Applied Probability and Queues
BAILEY • The Elements of Stochastic Processes with Applications to the Natural Sciences
BARNETT • Interpreting Multivariate Data
BARNETT and LEWIS • Outliers in Statistical Data, *Second Edition*
BARTHOLOMEW • Stochastic Models for Social Processes, *Third Edition*
BARTHOLOMEW and FORBES • Statistical Techniques for Manpower Planning
BATES and WATTS • Nonlinear Regression Analysis and Its Applications
BECK and ARNOLD • Parameter Estimation in Engineering and Science
BELSLEY, KUH, and WELSCH • Regression Diagnostics: Identifying Influential Data and Sources of Collinearity
BHAT • Elements of Applied Stochastic Processes, *Second Edition*
BLOOMFIELD • Fourier Analysis of Time Series: An Introduction
BOLLEN • Structural Equations with Latent Variables
BOX • R. A. Fisher, The Life of a Scientist
BOX and DRAPER • Empirical Model-Building and Response Surfaces
BOX and DRAPER • Evolutionary Operation: A Statistical Method for Process Improvement

Applied Probability and Statistics (Continued)

BOX, HUNTER, and HUNTER • Statistics for Experimenters: An Introduction to Design, Data Analysis, and Model Building
BROWN and HOLLANDER • Statistics: A Biomedical Introduction
BUNKE and BUNKE • Statistical Inference in Linear Models, Volume I
BUNKE and BUNKE • Nonlinear Regression, Functional Relations and Robust Methods: Statistical Methods of Model Building
CHAMBERS • Computational Methods for Data Analysis
CHATTERJEE and HADI • Sensitivity Analysis in Linear Regression
CHATTERJEE and PRICE • Regression Analysis by Example
CHOW • Econometric Analysis by Control Methods
CLARKE and DISNEY • Probability and Random Processes: A First Course with Applications, *Second Edition*
COCHRAN • Sampling Techniques, *Third Edition*
COCHRAN and COX • Experimental Designs, *Second Edition*
CONOVER • Practical Nonparametric Statistics, *Second Edition*
CONOVER and IMAN • Introduction to Modern Business Statistics
CORNELL • Experiments with Mixtures: Designs, Models and The Analysis of Mixture Data
COX • Planning of Experiments
COX • A Handbook of Introductory Statistical Methods
DANIEL • Biostatistics: A Foundation for Analysis in the Health Sciences, *Fourth Edition*
DANIEL • Applications of Statistics to Industrial Experimentation
DANIEL and WOOD • Fitting Equations to Data: Computer Analysis of Multifactor Data, *Second Edition*
DAVID • Order Statistics, *Second Edition*
DAVISON • Multidimensional Scaling
DEGROOT, FIENBERG and KADANE • Statistics and the Law
DEMING • Sample Design in Business Research
DILLON and GOLDSTEIN • Multivariate Analysis: Methods and Applications
DODGE • Analysis of Experiments with Missing Data
DODGE and ROMIG • Sampling Inspection Tables, *Second Edition*
DOWDY and WEARDEN • Statistics for Research
DRAPER and SMITH • Applied Regression Analysis, *Second Edition*
DUNN • Basic Statistics: A Primer for the Biomedical Sciences, *Second Edition*
DUNN and CLARK • Applied Statistics: Analysis of Variance and Regression, *Second Edition*
ELANDT-JOHNSON and JOHNSON • Survival Models and Data Analysis
FLEISS • Statistical Methods for Rates and Proportions, *Second Edition*
FLEISS • The Design and Analysis of Clinical Experiments
FLURY • Common Principal Components and Related Multivariate Models
FOX • Linear Statistical Models and Related Methods
FRANKEN, KÖNIG, ARNDT, and SCHMIDT • Queues and Point Processes
GALLANT • Nonlinear Statistical Models
GIBBONS, OLKIN, and SOBEL • Selecting and Ordering Populations: A New Statistical Methodology
GNANADESIKAN • Methods for Statistical Data Analysis of Multivariate Observations
GREENBERG and WEBSTER • Advanced Econometrics: A Bridge to the Literature
GROSS and HARRIS • Fundamentals of Queueing Theory, *Second Edition*
GROVES, BIEMER, LYBERG, MASSEY, NICHOLLS, and WAKSBERG • Telephone Survey Methodology
GUPTA and PANCHAPAKESAN • Multiple Decision Procedures: Theory and Methodology of Selecting and Ranking Populations
GUTTMAN, WILKS, and HUNTER • Introductory Engineering Statistics, *Third Edition*
HAHN and SHAPIRO • Statistical Models in Engineering

Bayesian Statistics:
Principles, Models, and Applications

Rev. T. Bayes
(1702–1761)

Bayesian Statistics: Principles, Models, and Applications

S. JAMES PRESS

Department of Statistics
University of California
Riverside, California

WILEY

JOHN WILEY & SONS

New York • Chichester • Brisbane • Toronto • Singapore

Library of Congress Cataloging-in-Publication Data

Press, S. James.
Bayesian statistics: principles, models, and applications/S. James Press.
p. cm.—(Wiley series in probability and mathematical statistics. Probability and mathematical statistics, ISSN 0271-6232)
Bibliography: p.
Includes indexes.
ISBN 0-471-63729-7
1. Bayesian statistical decision theory. I. Title. II. Series.

QA279.5.P75 1988 88-5407
519.5′44—dc19 CIP

Printed in the United States of America

10 9 8 7 6 5 4 3 2

To
Daryl, Jamie, and Julie

Preface

This book is intended to be an introduction to Bayesian statistics for students and research workers who have already been exposed to a good preliminary statistics and probability course from a classical (frequentist) point of view but who have had minimal exposure to Bayesian theory and methods. We assume a mathematical level of sophistication that includes a good calculus course and some matrix algebra but nothing beyond that. We also assume that our audience includes those who are interested in using Bayesian methods to model real problems in the various scientific disciplines. Such people usually want to understand enough of the foundational principles so that they will (1) feel comfortable using the procedures, (2) have no compunction about recommending solutions based upon these procedures to decision makers, and (3) be intrigued enough to go to referenced sources to seek additional background and understanding. For this reason we have tried to maximize interpretation of theory and have minimized our dependence upon proof of theorems.

The book is organized into two parts of four chapters each; in addition, the back of the book contains appendixes, a bibliography, and separate author and subject indexes. The first part of the book is devoted to theory; the second part is devoted to models and applications. The appendixes provide some biographical material about Thomas Bayes, along with a reproduction of Bayes' original essay.

Chapter I shows that statistical inference and decision making from a Bayesian point of view is based upon a logical, self-consistent system of axioms; it also shows that violation of the guiding principles will lead to "incoherent" behavior, that is, behavior that would lead to economically unsound decisions in a risky situation.

Chapter II covers the basic principles of the subject. Bayes' theorem is presented for both discrete and absolutely continuous random variables.

We discuss Bayesian estimation, hypothesis testing, and decision theory. It is here that we introduce prior distributions, Bayes' factors, the important theorem of de Finetti, the likelihood principle, and predictive distributions.

Chapter III includes various methods for approximating the sometimes complicated posterior distributions that result from applications of the Bayesian paradigm. We present large-sample theory results as well as Laplacian types of approximations of integrals (representing posterior densities). We will show how *importance sampling* as well as *simulation* of distributions can be used for approximation of posterior densities when the dimensions are large. We will also provide a convenient up-to-date summary of the latest Bayesian computer software available for implementation.

Chapter IV shows how prior distributions can be assessed subjectively using a group of experts. The methodology is applied to the problem of using a group of experts on strategic policy to assess a multivariate prior distribution for the probability of nuclear war during the decade of the 1980s.

Chapter V is concerned with Bayesian inference in both the univariate and multivariate regression models. Here we use vague prior distributions, and we apply the notion of predictive distributions to predicting future observations in regression models.

Chapter VI continues discussion of the general linear model begun in Chapter V, only here we show how to carry out Bayesian analysis of variance and covariance in the multivariate case. We will invoke the de Finetti notion of exchangeability (of the population mean vector distributions).

Chapter VII is devoted to the theory and application of Bayesian classification and discrimination procedures. The methodology is illustrated by applying it to the sample survey problem of second guessing "undecided" respondents.

Chapter VIII presents a case study of how disputed authorship of some of the Federalist papers was resolved by means of a Bayesian analysis.

The book is easily adapted to a one- or two-quarter sequence or to a one-semester, senior level, or graduate course in Bayesian statistics. The first two chapters and the appendixes could easily fill the first quarter, with Chapters III–VIII devoted to the second quarter. In a one-quarter or one-semester course, certain sections or chapters would need to be deleted; which chapters or sections to delete would depend upon the interests of the students and teacher in terms of the balance desired between (1) theory and (2) models and applications.

The book represents an expansion of a series of lectures presented in South Australia in July 1984 at the University of Adelaide. These lectures were jointly sponsored by the Commonwealth Scientific and Industrial Research Organization (CSIRO), Division of Mathematics and Statistics and by the University of Adelaide's Departments of Economics and Statistics. I am grateful to Drs. Graham Constantine, William Davis, and Terry Speed, all of CSIRO, for their stimulating comments on the original lecture material, for their encouragement and support, and for planting the seeds from which this monograph grew. I am grateful to Dr. John Darroch, Dr. Alastair Fischer, Dr. Alan James, Dr. W. N. Venables, and to other participants of the lecture series for their stimulating questions that helped to put the book into perspective. Dr. John Pratt and Dr. S. L. Zabell helped to clarify the issues about de Finetti's theorem in Section 2.9.3, and Dr. S. K. Sinha suggested an example used in Section 2.7.1. Dr. Persi Diaconis and Dr. Richard Jeffrey presented stimulating discussions about randomness, exchangeability, and some of the foundational issues of the subject in a seminar at Stanford University during winter quarter of 1984–1985, a sabbatical year the author spent visiting Stanford University. I am deeply grateful to Drs. Harry Roberts and Arnold Zellner for exposing me to Bayesian ideas. Dr. Stephen Fienberg provided encouragement and advice regarding publishing the manuscript. I am also grateful to Dr. Stephen Fienberg, Dr. Ingram Olkin, and an anonymous publisher's referee for many helpful suggestions for improving the presentation. I am very grateful for suggestions made by Dr. Judith Tanur who read the entire manuscript; to Dr. Ruben Klein who read Chapters I and II; and to Drs. Frederick Mosteller and David Wallace who read Chapter VIII. I also wish to thank graduate students, James Bentley, David Guy, William Kemple, Thomas Lucas, and Hamid Namini whose questions about the material during class prompted me to revise and clarify various issues. Mrs. Peggy Franklin is to be congratulated for her outstanding typing ability and for her forbearance in seeing me through the many iterations that the manuscript underwent. We think we have eliminated most, if not all, errors in the book, but readers could help the author by calling any additional ones they find to his attention.

S. JAMES PRESS

January, 1989
Riverside, California

Contents

APPENDIXES

Bayesian Statistics:
Principles, Models, and Applications

PART 1

Theory

CHAPTER I

Foundations

1.1 INTRODUCTION

We begin our introduction to Bayesian statistics with a discussion of some fundamental notions associated with randomness and probability, so that our applications will be developed from a solid foundation. We begin with the concept of "randomness"; once it is clear what we mean by randomness, we introduce the concept of probability, as "degrees of randomness." We discuss various interpretations of probability, and we place greatest emphasis on subjective probability. We then compare several distinct axiom systems, finally focusing on the Rényi system of conditional probability. Finally, we discuss some biographical background of the Reverend Thomas Bayes; we also discuss the background behind his celebrated theorem. For the convenience of the reader, we have reproduced the original famous essay of Bayes, along with additional biographical material, in the Appendixes.

1.2 RANDOMNESS

What do we mean when we refer to a phenomenon as "random"? We generally mean that the outcome of the phenomenon (event, or experiment) cannot be predicted with certainty. Suppose there could be "superscientists," that is, people who know all the physics, chemistry, biology, sociology, and other sciences that currently exist; in planning some experiment, one such scientist tries to use all of his/her knowledge about the underlying phenomenon to predict the outcome. But if even the superscientist cannot predict the outcome with certainty, the experiment has a "random" outcome.

For example, a small coin may be tossed 10 times. A machine may toss the coin, or a person may toss it. In either case, different forces and torques will be applied to the coin each time it is tossed, causing it to fall "heads" or "tails" with some considerable uncertainty. Even if the experiment is carried out in a laboratory so that the coin is tossed in a controlled environment that is "close to" vacuum conditions, and even if, in addition, the tossing is carried out by an electromechanical device that is "highly controllable," conditions will still be sufficiently uncontrolled so that even a superphysicist will be unable to predict the outcome with certainty. Such a phenomenon is random. (Magicians can sometimes toss a large coin, and by tossing it in a particular way, with considerable practice, they can predetermine the outcome. It is very difficult to do this with small coins.)

The degree of predictability of outcome for many phenomena depends upon the level of knowledge of the individual doing the predicting. Moreover, many phenomena are best thought of as "random" from a practical standpoint, because predictability is so difficult; others should be thought of as random because predictability is beyond all human ability. For example, Y. Kaneda of the University of Tokyo calculated the value of π to 134,217,700 digits (a little over 10^8) in 1987. But will human ability ever permit calculation of π to 10^{1000} digits? Since there is no formula for predicting the next digit in the transcendental number expansion, the only way to determine the last digit in the expansion is to calculate the digit that comes just before it. Since our universe will exist for only a finite amount of time, it does not seem possible for any human being ever to have sufficient time to be able to find such a number. So the number is random.

In an experimental setting, there will always be randomness and an inherent dependency upon initial conditions. In the context of the behavior of non-linear dynamical systems (biological, chemical, physical, or social), any small error in the observational measurement of the initial state of the system (the initial conditions) leads to unpredictability of the trajectory of the system through time. In such a case, the behavior of the system is random, or "chaotic". Moreover, the mathematical curve describing the limiting, long term, trajectory of the chaotic system may be "fractal". For a more extensive discussion of the mathematical concept of chaotic behavior, see, e.g., Abraham and Shaw, 1983, Volumes 0–4; F. Moon, 1987; Berge, Pomeau and Vidal, 1987; and Thompson and Stewart, 1986; or Gleick, 1987 (for a breezy treatment); for a non-mathematical, but very entertaining, discussion of mathematical randomness, see McKean, 1987. For extensive discussions of fractal curves, and of their relationships to chaos and mathematical randomness, see, for example, Mandelbrot, 1977a, 1977b; Peitgen and Richter, 1986; and Peitgen and Saupe, 1988.

In the case of coin flipping, slightly different initial conditions give rise to quite different physical behavior of the coin, making its final outcome very difficult to predict. Merely observing a phenomenon and measuring it also changes the result in random ways. This occurs at both the "macro" and "micro" levels, as we shall see below. At the macro level, the resulting change can often be taken into account. At the smallest micro level, there is no way to compensate.

The altered state associated with a phenomenon, attributable to merely observing it at the smallest micro level, was first quantified by Werner Heisenberg (see Heisenberg, 1927; also see Joos, 1934, pp. 650–653; and Pagels, 1982, pp. 69–73). Heisenberg observed that when we attempt to study phenomena occurring at so microscopic a level that we are within the structure of a single atom, we arrive at a measurement barrier that is inherent in nature—a certain randomness or uncertainty that cannot be eliminated by improving the measurement technique. Thus, if we try to measure the position x and momentum p of a subatomic particle simultaneously, we find we cannot ascertain initial conditions perfectly.

Let σ_X and σ_p denote the standard deviations of the error distributions of the measurements of (1) the X-coordinate of the position of a particle and (2) the X-coordinate of its momentum p, respectively. Heisenberg showed (in a quantum mechanical context) that we must always satisfy the "uncertainty principle" inequality,

$$\sigma_X \sigma_p \geq \hbar$$

where $\hbar \equiv h/2\pi = 1.054 \times 10^{-27}$ erg-sec, and h denotes a well-established constant of nature called *Planck's constant*. Thus, if by improving the experimental design and apparatus, σ_p is systematically reduced toward zero, σ_x will steadily increase to infinity, implying that to specify the momentum of the particle precisely (with zero error) is to lose all knowledge of the position of the particle at the same time. We note that although this effect is unnoticeable if we are measuring and predicting the positions of heavenly bodies or other material objects, for subatomic particles the effect is of fundamental importance. The problem here is that any means of observation by which atomic processes are conveyed to our perceptions causes a disturbance of the electron system of the atom. "It is meaningless to talk about the physical properties of objects without precisely specifying the experimental arrangement by which you intend to measure them" (Pagels, 1982, p. 76). This notion gives rise to our statistical philosophy of predictivism (see Section 2.9) as well as to our notion that prior probability distributions should incorporate information about the experiment that produces the observables (see "g-priors," in Section 2.7.3).

The altering of a phenomenon by merely observing it was described above for phenomena in the physical sciences where the phenomena are measured and observed at the "micro" level, that is, for measurements involving subatomic regions. Such alterations are explainable on the basis of the Heisenberg uncertainty principle.

On the "macro" level of observation there are other principles that tend to generate similar alterations that are found in other sciences. In the social sciences, for example, there is the "Hawthorne effect," and in the biomedical sciences there is the "placebo effect." These effects are explained below.

The Hawthorne effect (see Roethlisberger and Dickson, 1939) is sometimes observed in experiments on human subjects in which the people tend to behave differently simply because they are very aware of the fact that they are participants in an experiment and therefore they are being observed. In such cases, the subjects themselves, rather than the "treatments," induce the result being sought. The name of this effect derives from experiments carried out at the Hawthorne works of Western Electric Company outside of Chicago in a factory, in which it was desired to determine whether worker productivity would increase (decrease) by playing music in the factory during working hours and by changing other working conditions. Workers were asked to cooperate in the experiment, and it was observed that the productivity of workers increased substantially (almost regardless of which intervention took place). (Recent reexamination of the data in these experiments suggests that there might possibly be explanations other than the eponymous effect for the results found. Nevertheless, the effect does exist in some other experiments.) Thus, once a subject knows he/she is being experimented upon, there is an inherent uncertainty associated with the outcome of the experiment: Should the observed result be attributed to the treatment or to the Hawthorne effect?

The "placebo effect" is well known to workers in the biomedical sciences. A human subject is administered a placebo (a substance having no medicinal value) and is told that this drug will improve his/her state of health. The placebo operates psychologically on the individual, whose health then does improve in many instances. The individual believes strongly that the placebo will help, and so it does, but it is currently difficult to explain how the psyche can induce physiological change in this regard. The end result, at a macro level, however, is that once the subject knows he/she is being experimented upon, a certain inherent randomness in the outcome of the experiment arises. (Note that experiments carried out are also double "blind" in that neither the subject nor the person making diagnoses knows which subjects received the drug and which received the placebo.) If a patient is given a drug with true medicinal value, how much of the observed outcome is attributable to the drug, and how much to the placebo effect, so that an improvement would have been observed with even a placebo? To measure this difference we do "controlled experiments," and we give placebos to the control group. This uncertainty of outcome of the experiment is an inherent randomness at the macro level.

We must obviously be able to deal not only with randomness but also with degrees of randomness. The phenomenon of having "measurable rain" on a particular day in August in Los Angeles, California is random, but anyone who has examined historical records about rainfall in August in Los Angeles knows that rain occurs only rarely.

We all know that most coins we have ever seen tend to fall "heads" or "tails" with approximately equal frequency in a tossing experiment. (They don't have exactly equal frequency because the features carved on each side of the coin to establish identification of the "faces" cause a weight distribution imbalance which, in turn, generates a tendency for the coin to fall on either of its two faces with unequal frequency, but it is still close enough to equal frequency for most practical purposes.)

How should the different degrees of randomness that usually occur with random phenomena be specified or quantified? Conventional wisdom on this question suggests we use the notion of "probability."

1.3 PROBABILITY

1.3.1 Axiom Systems

Suppose in our coin tossing experiment the coin falls "heads" on each of 10 consecutive tosses. If we don't have prior knowledge of how the coin was fabricated and don't know how the coin has fallen in the past, what are we to believe about the bias of the coin? We consider two possibilities, stated as propositions:

A. The coin is strongly biased toward falling "heads."
B. The events of falling "heads" or "tails" are equiprobable, but we happen to have observed a possible, but unusual, run of "heads."

Should we believe A or B? Most nonstatisticians would be inclined to believe A, as would many statisticians; some statisticians would retain a belief of B. But if we change the level of occurrence from "all heads on 10 consecutive tosses" to "all heads out of 20 tosses," an even greater number of observers would accept proposition A instead of B. In fact, there is no way to know with certainty how this coin will fall; moreover, even if we were to toss it thousands of times and it always fell "heads," we could only gather an increasing amount of empirical evidence for more "heads" than "tails" by continuing to toss the coin more and more times. This accumulating empirical evidence affects our degree of belief about how the coin is likely to fall on the next, as yet unobserved, trial. If you wish to calculate the chance of getting say, 10 heads out of, say 10 trials, without knowing the bias of the coin, you may use de Finetti's theorem (see Sect. 2.9.3).

The outcome of an experiment, such as, how a coin falls (heads or tails), or any proposition, such as, "the Dodgers baseball team will win tomorrow's game," is called an "event." Your "degree of belief" about an event is called your *probability* of that event. [It is sometimes called your *subjective* (or personal) *probability* of that event, but we will not always make that distinction.] This probability of events may be assigned numerical values, as explained below.

The probability of an event A, expressed $P\{A\}$, or $P(A)$, is a non-negative real-valued function of events satisfying certain axioms that permit algebraic manipulation of probabilities. Various axiom systems have been proposed for the algebra of the events. Moreover, there have been several interpretations of $P(A)$, alternative to subjective probability (frequency probability; long run probability, defined as a limit; and logical probability, Keynes, 1921).

Kolmogoroff (1933) introduced the axiom system:

1. $0 \leq P\{A\}$.

2. $P\{\text{all possible events}\} = 1$.

3. Countable additivity: If $A_1, \ldots, A_n, \ldots$ are mutually exclusive and exhaustive, then $P\{\cup_{j=1}^{\infty} A_j\} = \sum_{j=1}^{\infty} P\{A_j\}$.

Probability functions satisfying this axiom system are called *mathematical probabilities* rather than subjective probabilities. De Finetti (1974) suggested that the Kolmogoroff system be modified so that Property 3 above need hold only for a finite collection of events (this is called the *finite additivity axiom*), and he required that the probability function $P(\cdot)$ be interpreted subjectively, as a personal probability for a given individual. The first formal mention of personal probability (see Sect. 1.3.3 for an operational definition) was by Frank Ramsey (1926). Kolmogoroff seems to have had a more objective interpretation in mind in which the interpretation of $P(\cdot)$ did not depend upon the individual.

De Groot (1970) developed an axiomatic system of probability in which the probability function was to be interpreted subjectively, but the concept of degree of belief was operationalized by using the notion of *relative likelihood*. Thus, $P(A) \leq P(B)$ means that, for an individual, event A is less likely (or equally likely) to occur than event B. The axiom system adopts the countable additivity property. (This is discussed further below.)

Ramsey (1926), Savage (1954), and de Finetti (1974) all favored the finite additivity axiom system. Countable additivity implies that the probability function $P(\cdot)$ is continuous, whereas finite additivity does not imply continuity. On the other hand, in the countable additivity system, some events (propositions) cannot be assigned a probability (these are called *nonmeasureable events*), whereas in a finite additivity axiom system, every event can be assigned a probability. In a countable additivity system, at every point of discontinuity x_0, of some cumulative distribution function $F(x)$, we have

$$\lim_{x \to x_0+} F(x) - F(x_0).$$

In a finite additivity system, this continuity result will no longer hold true. (Note that the cumulative distribution function, $F(x)$, is defined as $F(x) \equiv P\{X \leq x\}$. That is, it is the probability of the event $\{X \leq x\}$, where X denotes some random variable, which is itself a function which maps events into the real line.) Such a technical difficulty causes problems in developing asymptotic theory, for example. Since we never really pass all the way to the limit in applications, however, there is no difficulty in practice.

L. J. Savage (1954) expanded upon the finite additivity axiom system of subjective probability to develop a more broadly based seven-axiom system that could serve as the basis for a theory of rational decision making, using Bayesian principles (and he added in the "sure thing principle" as well, see Complement to Chapter 1). The Savage theory not only includes a finite additivity system of probability but also includes seven axioms appropriate for the mathematical scaling of preferences, axioms governing the use of lotteries to assess subjective probabilities of a decision maker. Savage's pioneering efforts in decision making were influenced by the earlier work of von Neumann and Morgenstern (1947), who developed a theory of computing economic gain and loss when engaging in a gamble. This "theory of utility" they developed became an integral component of Savage's theory. (Utility functions are defined and discussed in the Complement to this chapter, and in Sect. 2.3.1.) In fact, a major implication of Savage's work is that decisions should be made so as to maximize expected utility, conditional on all data already available (see Complement to Chapter 1, and Sect. 2.3.1). That is, optimal decision-making behavior results from maximizing average utility of the decision maker (i.e., averaged over the posterior distribution) (see Section 2.3.1, where we use loss functions, instead of utility functions, but they are just linearly related). The seven Savage axioms, as well as his theory of maximization of expected utility, are not repeated here (see Savage, 1954 and the Complement to Chapter 1 where the axioms are enumerated) because to do so would detract from the thrust of our effort. The Savage theory is normative, in that it is a theory for how people *should* behave, to be logical and rational, instead of a theory for how they actually do behave (an empirical theory).

Another formal axiomization of subjective probability, as it relates to utility and the scaling of preferences of a decision maker for various consequences of a lottery, was developed by Pratt, Raiffa, and Schlaifer (1964). The axiom system is logically equivalent to that of Savage (1954) but has, perhaps, more intuitive appeal and is more readily operationalizable. Related work, relying upon utility theory as it relates to subjective probability, may be found in Anscombe and Aumann (1963).

Rényi Axiom System

Rényi (1970) developed a countable additivity axiom system based upon conditioning. That is, Rényi assumed that all probabilities must be conditioned on certain prior historical information or evidence and that the entire axiom system should be erected on such a conditional foundation. A decision-making (or judgment-formulating) behavior conditioned on prior information seems closer to real-world behavior than an unconditional system. The Kolmogoroff axiom system is a special case. Let there be a set Ω called the space of elementary events, and let $\mathscr{A}$ denote a sigma-algebra of subsets of Ω. The elements $A, B, \ldots$ of $\mathscr{A}$ are called events. Let $\mathscr{B}$ denote a non-empty system of sets such that $\mathscr{B} \subseteq \mathscr{A}$. $P\{A \mid B\}$ will be defined for $A \in \mathscr{A}$, $B \in \mathscr{B}$. The Rényi axiom system (Rényi, 1970, p. 70) is given by the following:

1. For any events A, B, we have $P\{A \mid B\} \geq 0$, and $P\{B \mid B\} = 1$.
2. For disjoint events $A_1, \ldots$ (events that cannot occur in the same experiment) and some event B, we have

$$P\left\{\bigcup_1^\infty A_i \mid B\right\} = \sum_1^\infty P\{A_i \mid B\}.$$

3. For every event collection (A, B, C), $B \subseteq C$, $P\{B|C\} > 0$, we have

$$P\{A|B\} = \frac{P\{A \cap B|C\}}{P\{B|C\}}.$$

Note that if $P\{A|B\}$ doesn't depend upon B, A and B are said to be *independent*.

Thus, in the system we are adopting, continuity is subsumed. In this book, we adopt the Rényi conditional probability axiom system, and we interpret the probability function $P(A)$ subjectively, that is, as a degree of belief about the event A. Our rationale for adopting the Rényi system is pragmatic: real world judgment formulation and decision making is almost invariably conditional upon prior information and experience. Moreover, at the minor cost of introducing an axiom system with non-measurable events, the countable additivity axiom of the Rényi system permits us to have continuity and asymptotic theory as part of our armamentarium.

It will be seen later that for some purposes it will be useful to use probability distributions that spread their masses uniformly over the entire real line. In all of the axiom systems mentioned, such continuous distributions are "improper," because their densities do not integrate to unity. We will find that a good use for such improper distributions is to represent an individual's distribution of belief about an unknown quantity that can lie anywhere on the real line, prior to observing the outcome of an experiment. Suppose the individual to be "you." If all possible values of the unknown quantity seem to you to be equally likely over a certain interval, you would adopt a uniform distribution over that interval. But how large should the interval be if you are really not knowledgeable about the quantity? As the length of the interval increases to infinity, to reflect a greater degree of ignorance or uncertainty, the distribution goes from proper to improper. See Section 2.7.2 for a detailed discussion of uniform (vague) prior distributions and for an explanation of why they are of such fundamental importance in Bayesian statistics. For our purposes it is useful to appreciate that the Rényi axiom system of probability accommodates (by means of limits of ratios) probabilities on the entire real line and that it is thus closest to the system of probability required for Bayesian inference. It will be seen that we will have frequent occasion to use posterior distributions which are proper but which have resulted, by application of Bayes theorem, from use of improper prior distributions. The mathematical formalism required to handle probability calculations involving improper distributions in a rigorous way can be accommodated by the Rényi probability structure but cannot be accommodated very well by any of the other systems. It would be useful to have a formal axiomatic theory of subjective probability which is not only conditional in the sense of Rényi [so that it accommodates probabilities on the entire real line (unbounded measures)] but which also captures the advantages of a finitely additive probability system. Some 30 distinct axiom systems, which are mostly small variations on one another, which are attempts to weaken the axioms, are reviewed by Fishburn, 1981. Such systems include probability axioms as well as preference ordering axioms. For an up-to-date tour through the various axiom systems, see Fishburn, 1986.

1.3.2 Coherence

An individual whose probability statements (beliefs) about a collection of events do not satisfy our (Rényi) system of axioms (for example, if they are mutually inconsistent) is said to be "incoherent" (see de Finetti, 1937, p. 111). (According to cognitive psychologists, observed human behavior tends to be incoherent, whereas our axiom systems and normative interpretation specify "ideal" appropriate behavior.) If an incoherent individual is willing to make a sequence of bets using these (incoherent) probabilities, and he/she considers each bet fair or favorable, he/she will suffer a net loss no matter what happens. (Such a bet is called a *Dutch book*.)

For example, suppose you claim that your degree of belief about Event A is $P\{A\} = 0.7$. You claim moreover that your degree of belief about Event B is $P\{B\} = 0.5$. You furthermore know that A and B are independent, and you believe $P\{\text{both } A \text{ and } B\} = 0.1$. In such a case you are being incoherent because

$$P\{\text{both } A \text{ and } B\} = P\{A\}P\{B\} = (0.7)(0.5) = 0.35.$$

Whenever multiple subjective probabilities are assessed on the same individual, coherency checks should be run to preclude the possibility of Dutch book, and to ensure internal consistency.

In another example, suppose you feel $P\{A\} = 0.75$, and also $P\{\bar{A}\} = 0.75$, where $\bar{A}$ denotes the complementary event to A. We define the two lotteries: Lottery (1)—N_1 bets \$3 on event A, and N_2 bets \$1 on event $\bar{A}$; and Lottery (2)—N_3 bets \$3 on event $\bar{A}$, and N_4 bets \$1 on event A. You would feel both lotteries are fair, and you would be willing therefore to take either side of both lotteries. Suppose you choose N_1 and N_3, and I choose N_2 and N_4. If A occurs you receive \$1 from Lottery (1) but must pay \$3 from Lottery (2), with a net loss of \$2. If $\bar{A}$ occurs, you must pay \$3 in

Lottery (1), but collect \$1 in Lottery (2), for a net loss again of \$2. So you lose \$2 no matter what happens—so I have made a Dutch book against you because you were incoherent.

1.3.3 Operationalizing Subjective Probability Beliefs

How should an individual think about an event (Event A) to quantify his/her judgment about how strongly he/she believes the event is likely to occur? Equivalently, how does an individual generate $P(A)$? One way to do it which is particularly useful for events of extremely small or large probability is to use De Groot's approach of specifying "relative likelihoods" of events (see Section 1.3.1). Individuals can often provide ordinal rankings for the likelihoods of events, even though they may find it much more difficult to specify absolute numerical probabilities for specific events (especially when the events are either quite remote or extremely likely). In Section 4.7, for example, it will be seen that subjects were asked to specify ordinal rankings for their subjective probabilities of nuclear war in the 1980s.

Another approach to operationalizing probability statements, and the one we recommend for most frequent use, is to operationalize degree of belief in terms of lotteries. [This idea is attributed to Ramsey (1926), who required the notion of an "ethically neutral proposition" called $\tilde{p}$, for which degree of belief is 1/2. Your degree of belief in $\tilde{p}$ is 1/2 if you are indifferent to receiving either (1) α if $\tilde{p}$ is true and β if $\tilde{p}$ is false or (2) α if $\tilde{p}$ is false and β if $\tilde{p}$ is true.]

Suppose you are interested in some event, say Event A. Denote by $p \equiv P\{A\}$ your degree of belief that Event A will occur. Your degree of belief is illustrated as follows. Suppose you are offered a choice between the following two options:

1. receiving some small reward R (such as \$1) if Event A occurs but receiving no reward if A does not occur;
2. engaging in a lottery in which you win a small reward R with probability p, but you lose and receive no reward with probability $(1 - p)$.

If you are indifferent between choices 1 and 2, then your degree of belief in Event A occurring is p. The size of the reward R plays no role as long as it is small. In the remainder of this book we assume that when you say your degree of belief in Event A is p, you mean you are indifferent to the pair of choices given above.

For example, suppose the local weather forecaster claims the probability of measurable precipitation in your geographical area for tomorrow is 0.7, or 70%. The forecaster should then be indifferent between receiving a small reward R if precipitation occurs or receiving nothing if precipitation does not occur, as compared with receiving R if a white ball is selected randomly from an urn containing exactly seven white balls and three nonwhite balls or receiving nothing otherwise. Note that this definition of probability affords us the opportunity to score a probability assessor on how well he/she assesses. Therefore, if measurable precipitation actually occurs, we can give the weather forecaster positive feedback and a reward. If it does not occur, he/she gets negative feedback and no reward. If the forecasting is carried out repeatedly the forecaster should improve his/her probability assessing ability.

Scoring becomes particularly important to society when the subjective probability assessment involves the chance of rare events (one time in one million, one time in one billion, etc.) occurring (nuclear power plant disasters, etc.). In this context it has been found that simultaneously competing risks can often confuse the assessor, and result in much lower "perceived risks" than those that actually occur (see Freudenburg, 1988).

As another example, suppose you believe that a certain coin falls "heads" or "tails" with equal probability, and I offer you a lottery ticket on the toss of that coin that pays \$1 if "heads" comes up but pays nothing otherwise; then you should be willing to pay half a dollar (the expected return) for the ticket, and if I accept that price you should feel the bet is "fair."

These notions operationalize our way of thinking about subjective probabilities.

1.3.4 Comparing Probability Definitions

How is our degree-of-belief notion of probability to be reconciled with other interpretations of probability? The notion of probability as a "long-run frequency" (see von Mises, 1957, 1964) is a theoretical construct with no obvious way to be operationalized in a finite number of trials. To clarify this point, our coin-tossing example will serve well again. Every time an experimenter tosses the same coin the initial conditions change. So his/her belief about "heads" coming up on this toss should depend upon the initial conditions of the toss. I have never personally tossed a coin as many as one million times (nor do I know of anyone else who has), so I don't know what would happen after even that few a number of tosses, but I doubt that the proportion of "heads" found would be precisely 0.5 for most any normal coin. In spite of these deficiencies of the construct, long-run frequency remains a useful notion for some applications, as long as we remain prepared to recognize that the notion must be abandoned if it contradicts our degree of belief (in small samples in any given situation).

Using the Kolmogoroff concept of mathematical probability, if we toss a coin 20 times and obtain 20 "heads," we should continue to believe the coin is "fair," if that is what we believed prior to the first toss. We should believe this in spite of the fact that a fair coin could only fall that way with a mathematical probability of about one in a million (actually, one chance in 1,048,576). Here, "fair" means its two faces are equiprobable in the long-run frequency sense. In any real-world situation, however, such as someone in a gaming situation in a casino, the individual is unlikely to believe the 20 "heads" occurred with probability 0.5 on each trial and is much more likely to modify his/her belief about the fairness of the coin. (Individuals will, of course, vary according to how many straight successes it will take before they will alter their belief about the bias of the coin.) (The casino game of roulette almost perfectly illustrates this phenomenon. If you bet on "Red" for 20 successive turns of the wheel, and "Black" eventuates each time, you are unlikely to believe the wheel is fair.)

In the case of the coin, if you are willing to accept as "prior knowledge" (prior to the tossing of the coin) the knowledge that the coin has a particular bias (such as equiprobable outcomes), then our predictions of outcomes coincide with those of Kolmogoroff (mathematical probability). For philosophical elaborations of the various definitions and interpretations of probability, such as that of Laplace (1814), as compared with probability as a long-run frequency, logical probability, and subjectivist probability; and for further discussion of the construction of logical axiomatic systems of scientific inference; and arguments for and against the finite, versus countable, additivity axiom; etc., see Carnap and Jeffrey (1971), Good (1983) (a collection of some of his more philosophical papers), Jeffrey (1980) (Vol. 2 of Carnap and Jeffrey, 1971), Keynes (1921), Lindley (1976), Popper (1968) and Skyrms (1984).

1.4 THOMAS BAYES

Thomas Bayes was a Presbyterian minister and mathematician who lived in England in the 1700s (born circa 1702 and died April 17, 1761). Richard Price, interested in Bayes' research, submitted Bayes' manuscript on inverse probability in the binomial distribution to the professional journal, *Philosophical Transactions of the Royal Society*, which published the paper (posthumously) in 1763, an article reproduced in Appendix 4 of this book, along with biographical information reproduced in Appendices 1–3.

There has been some mystery associated with Thomas Bayes. We are not quite certain about the year of his birth. Moreover, questions have been raised about who actually wrote the paper generally attributed to him (see Stigler, 1983) and about what his theorem actually says (see Stigler, 1982). Common interpretation today is that the paper proposed a method for making probability inferences about the parameter of a binomial distribution conditional on some observations from that distribution. (The theorem attributed to Thomas Bayes is given in Proposition 9, Appendix 4; the Scholium which follows it has been controversial, but has been widely interpreted to mean that "knowing nothing" about a parameter implies we should take a uniform distribution on it.) Common belief is that Bayes assumed that the parameter had a uniform distribution on the unit interval. His proposed method for making inferences about the binomial parameter is now called *Bayes' theorem* (see Section 2.2) and has been generalized to be applicable beyond the binomial distribution, to any sampling distribution (Bayes appears to have recognized the generality of his result but elected to present it in that restricted binomial form regardless). It was Laplace (1774) who stated the theorem on inverse probability in general form, and who, according to Stigler (1986), probably never saw Bayes' essay, and probably discovered the theorem independently. (Bayes carried out his work in England, where his theorem was largely ignored for over 20 years; Laplace carried his work in France.) Jeffreys (1939) rediscovered Laplace's work.

Your distribution for the parameter is called your *prior distribution*, because it represents the distribution of your degree of belief about the parameter prior to your observing any data, i.e., prior to your carrying out any experiment that might bear on the value of the parameter. (Bayes assumed a uniform prior distribution for the success probability in a binomial distribution.) Bayes' theorem gives a mathematical procedure for updating your prior belief about the value of the parameter to produce a *posterior distribution* for the parameter, one determined subsequent to your having observed the outcome of an experiment bearing on the value of the unknown parameter. Thus, Bayes' theorem provides a vehicle for changing, or updating, the degree of belief about a parameter (or a proposition) in light of more recent information. It is a formal procedure for merging knowledge obtained from experience, or theoretical understanding of a random process, with observational data. Thus, it is a normative theory for learning from experience. The theorem is given mathematical form, with examples, in Chapter II (Section 2.2).

The ideas in the theorem attributed to Bayes were really conceived earlier by James Bernoulli in 1713, in Book 4 of his famous treatise on probability, *Ars Conjectandi* ("The Art of Conjecturing"), published posthumously. In that book, James Bernoulli—or Jakob Bernoulli, as he was known in German—not only developed the binomial theorem and laid out the rules for permutations and combinations but also posed the problem of inverse probability of Bayes (who wrote his essay 50 years later); however, Bernoulli didn't give it mathematical structure. In *Ars Conjectandi*, James Bernoulli wrote:

> To illustrate this by an example, I suppose that without your knowledge there are concealed in an urn 3000 white pebbles and 2000 black pebbles, and in trying to determine the numbers of these pebbles you take out one pebble after another (each time replacing the pebble you have drawn before choosing the next, in order not to decrease the number of pebbles in the urn), and that you observe how often a white and how often a black pebble is withdrawn. The question is, can you do this so often that it becomes ten times, one hundred times, one thousand times, etc., more probable (that is, it be morally certain) that the numbers of whites and blacks chosen are in the same 3:2 ratio as the pebbles in the urn, rather than in any other different ratio? [in translation]

This is the problem of inverse probability that concerned 18th-century mathematicians (see Stigler, 1986, Chapter 2). According to Egon Pearson (1978, p. 223), James Bernoulli "...was destined by his father to be a theologian, and [he] devoted himself after taking his M.A. at Basel (Switzerland) to theology." This endeavor involved Bernoulli in philosophical and metaphysical questions (similar training was, of course, true for Bayes, who was a minister). In fact, Maistrov (1974, p. 67), evaluating Bernoulli, believes he was an advocate of "metaphysical determinism," a philosophy very similar to that of Laplace, some of whose work didn't appear until 100 years after Bernoulli. Writing in *Ars Conjectandi*, in the first chapter of the fourth part, Bernoulli said:

> For a given composition of the air and given masses, positions, directions, and speed of the winds, vapor, and clouds and also the laws of mechanics which govern all these interactions, tomorrow's weather will be no different from the way it should actually be. So these phenomena follow with no less regularity than the eclipses of heavenly bodies. It is, however, the usual practice to consider an eclipse as a regular event, while [considering] the fall of a die or tomorrow's weather as chance events. The reason for this is exclusively that succeeding actions in nature are not sufficiently well known. And even if they were known, our mathematical and physical knowledge is not sufficiently developed, and so, starting from initial causes, we cannot calculate these phenomena, while from the absolute principles of astronomy, eclipses can be precalculated and predicted... The chance depends mainly upon our knowledge. [in translation]

In the preceding excerpt, Bernoulli examined the state of tomorrow's weather, given today's observational data that relate to weather and a belief about tomorrow's weather. This is precisely the kind of question addressable by Bayes' theorem of 1763, in terms of degree of belief about tomorrow's weather, given today's observations and a prior belief about tomorrow's weather. (Moreover, the development of quantum mechanics and the Heisenberg uncertainty principle (Sect. 1.2) has negated Bernoulli's view of "chance.")

Additional background on the life of Thomas Bayes may be found in Pearson (1978).

1.5 SUMMARY

In this chapter we have tried to summarize the foundational issues upon which Bayesian statistics is based. We started with the elemental notion of randomness which is intrinsic to nature. We then developed our ideas about "probability" as a quantitative characterization of randomness. We compared the axiomatic developments of Kolmogoroff (mathematical probability) with those of De Groot, de Finetti, Ramsey, Rényi, and Savage (subjective probability). We discussed coherence as well as how to operationalize our degree of belief about an unknown quantity. We considered the long-run frequency interpretation of probability of von Mises, and finally we discussed the Reverend Thomas Bayes and the intuitive interpretation of his famous theorem. The mathematical treatment of the theorem is presented in Chapter II, along with other basic principles of Bayesian inference and decision making.

COMPLEMENT TO CHAPTER I: THE AXIOMATIC FOUNDATION OF DECISION MAKING OF LEONARD J. SAVAGE

Definition. **f** will be used to denote an act, that is, a function, attaching the consequence $f(s)$ to the event s.

Axiom 1. The relation $\leq\cdot$ is a simple ordering among acts. Thus, $\mathbf{f} \leq\cdot \mathbf{g}$ means that act **f** is not preferred to act **g**, or there is indifference between acts **f** and **g**. So if a person had to decide between acts **f** and **g**, and no other acts were available, he/she would decide upon **g**.

Axiom 2. For every act **f** and every act **g**, and every event B $\mathbf{f} \leq\cdot \mathbf{g}$ given B, or $\mathbf{g} \leq\cdot \mathbf{f}$ given B. Also, if $\mathbf{f} \leq\cdot \mathbf{g}$, and $\mathbf{g} \leq\cdot \mathbf{h}$, then $\mathbf{f} \leq\cdot \mathbf{h}$.

"Sure Thing Principle". If (**f** $\leq\cdot$ **g**)$|B$, and (**f** $\leq\cdot$ **g**)$|\bar{B}$, then **f** $\leq\cdot$ **g**. Here, $\bar{B}$ denotes the event: not B. That is, if your choice is the same regardless of event B, then B is irrelevant to your choice. The principle is often violated empirically. The most famous refutation is due to Allais (1953).

Definition. **f** $\equiv g$ means a constant act. That is, for every event, s, $f(s) = g =$ the consequence of the act, and act **f** results in the same consequence regardless of s.

Axiom 3. If $f(s) = g$, $f'(s) = g'$, for every $s \in B$, and B is not null, then **f** $\leq\cdot$ **f**$'$, given B, if and only if $g \leq g'$.

Definition. $A \leq^* B$ means that event A is not more probable than event B. (Note that "probable" means in a subjective probability sense.)

Axiom 4. For every A, B, $A \leq^* B$ or $B \leq^* A$.

Axiom 5. It is not true that for every f and f', $f \leq^* f'$. That is, there is at least one pair of consequences f, f' such that $f' < f$.

Axiom 6. If **g** $<\cdot$ **h**, and f is any consequence, then there exists a partition of the space of all events, S, such that, if **g** or **h** is so modified on any one element of the partition as to take the value f at every s there, other values being undisturbed, then the modified **g** remains less than **h**, or **g** remains less than the modified **h**, as the case may require.

Definition. **f** $\leq\cdot$ g, given B, if and only if **f** $\leq\cdot$ **h**, given B, when $h(s) = g$ for every s.

Axiom 7. If **f** $\leq\cdot$ $g(s)$, given B, for every $s \in B$, then **f** $\leq\cdot$ **g**, given B.

Conclusions

Savage (1954) shows that Axioms 1–7 imply the existence of a utility function $U(\mathbf{f})$, such that:

1. If **f** and **g** are bounded, then **f** $\leq\cdot$ **g** if and only if $U(\mathbf{f}) \leq U(\mathbf{g})$;
2. If **f** and **g** are bounded, and $P\{B\} > 0$, then **f** $\leq\cdot$ **g**, given B, if and only if $E\{[U(\mathbf{g}) - U(\mathbf{f})]|B\} \geq 0$.

Utility (see also Section 2.3.1)

A utility function $U(\mathbf{f})$ is a real valued function defined on the set of consequences, such that $\mathbf{f} \leq\cdot\, \mathbf{g}$ if and only if $U(\mathbf{f}) \leq U(\mathbf{g})$. Let $f(s) = g$. Then, to each consequence g there corresponds a numerical valued function $U(g)$ which represents the value you place on the consequence g. Moreover, any linear function $U^*(g) = aU(g) + b$, where $a > 0$, is also a utility function. In scaling utilities, for convenience we often define $U(g)$ on the unit interval, and we take the utility of the worst and best possible consequences in a gamble (decision problem) to be zero and unity, respectively.

In this book we adopt the Rényi axiom system for probability, with a subjective interpretation of the probability function. In addition, for decision making, we will adopt the principle of maximization of expected utility.

EXERCISES

1.1 What is the "Heisenberg uncertainty principle," and why is it important?

1.2* Give an explanation of the mathematical view of "chaos." (*Note*: This involves some research.)

1.3 Explain the meaning of the "Hawthorne effect," and explain why it is important.

1.4 What is meant by the "placebo effect," and why is it important?

1.5 Distinguish between finite and countable additivity, and explain the importance of the difference.

1.6 Explain the importance of the Rényi axiom system of probability for Bayesian inference.

1.7 What is meant by "coherence"?

1.8 What is meant by a "relative likelihood," and how might you use one?

1.9 What is the meaning of the (von Mises) long-run frequency approach to probability?

1.10* By going to the original reference sources, explain who Stigler believes actually wrote Bayes' theorem and also explain what Bayes' theorem was really saying.

*Asterisked exercises require reference to sources outside of this text. Full reference information can be found in the bibliography at the back of this book.

1.11 Explain how to operationalize the interpretation of the statement: "I believe that there is a 70% chance that the Los Angeles Dodgers baseball team will win the World Series next year."

1.12 Explain why a frequency interpretation of probability cannot possibly interpret the statement in Exercise 1.11.

1.13 Why is the book *Ars Conjectandi* by James Bernoulli of particular interest to people interested in Bayesian statistical inference?

1.14 What is meant when we say that a phenomenon has a random outcome?

1.15 Distinguish between prior and posterior distributions.

1.16 What is meant by "Savage's axioms," and what is the context to which this axiom system applies?

1.17 What is meant by utility theory?

1.18 Explain the assertion "mathematical probability is merely a special case of subjective probability."

1.19 How would you operationalize your assertion that your probability of rain tomorrow in Riverside, California is 0.7?

1.20 What is meant by a "normative theory of inference and decision making behavior" as compared with an empirical theory? (Hint: see Section 1.3.1.)

CHAPTER II

Principles

2.1 INTRODUCTION

This chapter presents the basic principles behind Bayesian inference. We begin with an example about a parts supplier whose boxes of parts contain unknown numbers of defectives. The problem is to use historical information about the numbers of defective parts the supplier has shipped in the past to help to determine the probability of his having defectives in the current shipment. We then introduce Bayes' theorem for discrete parameters, and we apply the theorem to the parts-supplier problem. We introduce Bayesian estimation in a decision-theoretic context, and we show how to test hypotheses from a Bayesian viewpoint (we also discuss the interesting Lindley paradox). Then we introduce Bayes' theorem for continuous parameters. Section 2.7 is devoted to the fundamental issues related to prior distributions. There we discuss how to express varying degrees of uncertainty about an unknown quantity, from the greatest uncertainty (vague priors) to small uncertainty, and we discuss the use of "convenience priors" (natural conjugate families). We discuss the fundamental likelihood principle (basing all inferences on the information in the likelihood function), predictive distributions, and predictivism. We discuss Bayesian interval estimation as well as robust and nonparametric Bayesian inference, including the Bayesian bootstrap. A brief overview of principles of univariate Bayesian analysis is given in Roberts, 1978; a similar brief overview of multivariate Bayesian analysis is given in Press, 1985b. Other reviews of Bayesian principles may be found in Lindley (1972, 1976) and in Zellner (1987).

2.2 BAYES' THEOREM FOR DISCRETE PARAMETERS

2.2.1 Example (Defective Parts)

A box containing eight parts is received from a supplier. In the past, 70% of all such boxes have had zero defective parts; 20% have had one defective part, and 10% have had two defective parts. We assume therefore that all boxes containing eight parts will have either 0, 1, or 2 defectives. Three parts are selected at random from the box of eight, and one part is found to be defective. What is the probability the box of eight parts received from the supplier actually contained two defective parts?

The answer is about 42%, and we will calculate this answer by using Bayes' theorem (see Section 2.4.2).

2.2.2 Bayes' Theorem (Discrete Parameter)

Let $\mathbf{X}_1, \ldots, \mathbf{X}_n$ denote independent and identically distributed observable random vector variables, each with either probability mass function $f(\mathbf{x}|\boldsymbol{\theta})$ or density (with respect to Lebesgue measure) $f(\mathbf{x}|\boldsymbol{\theta})$; that is, f denotes the mass function or density of a random vector $\mathbf{X}$, conditional upon another random variable $\boldsymbol{\Theta} = \boldsymbol{\theta}$. Θ is assumed to be unobservable, and $\boldsymbol{\theta}$ denotes the numerical vlaue at which Θ is conditioned. Assume $\boldsymbol{\Theta}$ is discrete (the case in which $\boldsymbol{\Theta}$ is continuous is treated in Sect. 2.5), and let $g(\boldsymbol{\theta})$ denote its probability mass function. (Since it is not likely to cause confusion, we will drop the $\boldsymbol{\Theta}$ and use $\boldsymbol{\theta}$ to denote both the random variable and its conditioned value.) Then, Bayes' theorem asserts that the probability mass function of $\boldsymbol{\theta}$, for given $(\mathbf{X}_1, \ldots, \mathbf{X}_N)$, is given by

$$h(\boldsymbol{\theta}|\mathbf{x}_1, \ldots, \mathbf{x}_N) = \frac{f(\mathbf{x}_1|\boldsymbol{\theta}) \cdots f(\mathbf{x}_N|\boldsymbol{\theta}) g(\boldsymbol{\theta})}{\sum_{\theta} f(\mathbf{x}_1|\boldsymbol{\theta}) \cdots f(\mathbf{x}_N|\boldsymbol{\theta}) g(\boldsymbol{\theta})}.$$

Proof: Trivial, using the definition of conditional probability. □

2.2.3 Remarks

1. Since the denominator depends only on the $\mathbf{x}_i$'s (and not on $\boldsymbol{\theta}$), we may write $h(\boldsymbol{\theta}|\mathbf{x}_1, \ldots, \mathbf{x}_N) \propto L(\mathbf{x}_1, \ldots, \mathbf{x}_N|\boldsymbol{\theta}) g(\boldsymbol{\theta})$, where $\propto$ denotes proportionality and where $L(\mathbf{x}_1, \ldots, \mathbf{x}_N|\boldsymbol{\theta}) \equiv f(\mathbf{x}_1|\boldsymbol{\theta}) \cdots f(\mathbf{x}_N|\boldsymbol{\theta})$ denotes the *likelihood function* of the data given the parameter $\boldsymbol{\theta}$, for independent data. Note that when viewed as a function of $\boldsymbol{\theta}$, the likelihood function is unique only up to a multiplicative constant. The constant makes no difference in Bayes' theorem since it is absorbed into the proportionality constant.

2. $g(\boldsymbol{\theta})$ is called the *prior* probability mass function of $\boldsymbol{\Theta}$, since it is determined prior to observing **X** in the current experiment; that is, $g(\boldsymbol{\theta})$ is based upon previous practical experience and understanding.

For example, if $\boldsymbol{\theta}$ denotes the mean value of a population, then $g(\boldsymbol{\theta})$ denotes the degree of belief you have about this average value, based upon earlier experience with populations like this one. For example, you may feel that your degree of uncertainty about the value of Θ may be expressed by a distribution in which all possible values of Θ are equally likely. We will discuss $g(\boldsymbol{\theta})$ in greater detail in Section 2.7.

3. $h(\boldsymbol{\theta}|\mathbf{x}_1, \ldots, \mathbf{x}_N)$ is called the *posterior* probability mass function (pmf) of $\boldsymbol{\theta}$, given the current data, since it is determined posterior to, or after, observing the current set of data. (It is the pmf of our updated belief distribution.)

4. Equivalent statement of Bayes' theorem:

$$\text{Posterior} \propto \text{Likelihood} \times \text{Prior}.$$

2.2.4 Interpretation of Bayes' Theorem

If we have *no* current observations to draw upon, then we must make all of our judgments about $\boldsymbol{\theta}$ from previous experience; that is, we use the prior probability mass function, $g(\boldsymbol{\theta})$, only.

If we have *both* previous experience and current perceptions based upon observational data, we revise $g(\boldsymbol{\theta})$ according to Bayes' theorem and base our inferences about $\boldsymbol{\theta}$ on $h(\boldsymbol{\theta}|\mathbf{x}_1, \ldots, \mathbf{x}_N)$, the posterior probability mass function.

2.3 BAYESIAN METHODS

2.3.1 Estimation: Decision Theory—Utility

When a decision is to be made among various possible actions or acts that can be taken in a given situation, we can associate with action a_i a consequence q_i, and q_i will result with probability p_i, $i = 1, \ldots, n$. The consequence q_i will be valued differently by different people. If q_i is a pair of first-row seats at a concert, a person who loves music will greatly value q_i, whereas someone who doesn't care for music will not be very interested in q_i. As a second example, if q_i represents \$100, a poor person will value the money more than a wealthy person. The value an individual places on q_i is referred to as his/her *utility* for that consequence. If you do not prefer q_i to q_j, then we write $q_i <\cdot\, q_j$. If you are indifferent between q_i and q_j, or you do not prefer q_i to q_j, we write $q_i \leq\cdot q_j$. We assume all consequences can be preferentially ordered (see Complement to Chapter 1).

Your utility function, $U(q)$, defined on the consequences in a decision problem, is a real-valued function satisfying the following properties:

1. If $q_1 <\cdot q_2$, then $U(q_1) < U(q_2)$; if you are indifferent between q_1 and q_2, then $U(q_1) = U(q_2)$.
2. If you are indifferent toward receiving q_1 with certainty or toward taking a gamble in which you receive q_2 with probability p and receive q_3 with probability $(1 - p)$, then

$$U(q_1) = pU(q_2) + (1 - p)U(q_3).$$

Thus, a gamble is equivalent to a probability distribution over the consequences.

It is sometimes argued that $U(q)$ should be a bounded function. Note also that utility functions are uniquely defined only up to a linear transformation, so that if $U(q)$ is a utility function, so is $aU(q) + b$, for $a > 0$. For this reason we usually define $U(q)$ on the unit interval, and we take $U(q_*) = 0$ and $U(q^*) = 1$, where q_* and q^* are the worst and best consequences that can occur in a given decision problem. Then, for all q, we have $0 \leq U(q) \leq 1$. For an additional discussion of utility, see the Complement to Chapter 1.

Now suppose the decision problem of interest involves consequences that are gambles (the earlier discussion is a special case). Let P_1 and P_2 denote two probability distributions over the consequences. Then it follows from the above properties of utilities that $P_1 \leq\cdot P_2$ if and only if $E[U(P_1)] \leq E[U(P_2)]$. Thus, in deciding among consequences that are gambles, you should prefer the one with the greatest expected utility. These assertions about expected utility follow from the Savage axioms (Complement to Chapter 1).

When utility functions are fitted to human preferences, utility as a function of money received in a gamble is generally a concave function, which reflects most individuals' aversion to risk. A person who likes to take risks will typically have a convex utility function, when utility is viewed as a function of money. Often, a utility function will be expanded in Taylor series about the expected consequence, $\bar{q}$, including up to second-order terms. Such an expansion gives rise to quadratic utility functions (which are unbounded in θ). Thus, $U(q) = U(\bar{q}) + (q - \bar{q})U'(\bar{q}) + \frac{1}{2}(q - \bar{q})^2U''(\bar{q}) + \cdots$.

It is sometimes more convenient to talk in terms of "loss functions" instead of utility functions. By loss we mean "opportunity loss," that is, the difference between the utility of the best consequence you could have obtained and the utility of the actual one you received. Thus, loss is always non-negative, by definition.

A point estimator of θ, a parameter of a distribution, is often taken to be the mean of the posterior distribution of θ. This is because the posterior mean minimizes the mean quadratic loss function (risk) in a decision-theoretic context. (For Bayesian discussions of decision theory see, e.g., Berger, 1980, 1985; De Groot, 1970; Ferguson, 1967; Lindley, 1985; Pratt, Raiffa, and Schlaifer, 1965; Press, 1982; and Raiffa and Schlaifer, 1961.) Equivalently, the posterior mean estimator maximizes the expected quadratic utility. Other loss functions would imply other Bayesian estimators (e.g., an absolute error loss function implies a posterior median estimator; for an unspecified loss function we sometimes use the posterior mode). For Bayes estimation with respect to asymmetric loss functions, see Varian, 1975, and Zellner, 1986.

For example, suppose $r|\theta, n$ denotes the number of successes in n trials in a binomial distribution with parameter θ, and we would like a Bayesian estimator of θ when the loss function is $L(\theta, \hat{\theta}) = (\theta - \hat{\theta})^2$, for some estimator $\hat{\theta}$. Suppose the posterior distribution for θ is discrete, with pmf $h(\theta|r, n)$, and the possible values for θ are $\theta_1, \ldots, \theta_k$. The expected loss (averaged over the posterior distribution) is

$$E\left[L(\theta, \hat{\theta})|r, n\right] = \sum_{i=1}^{k} \left(\theta_i - \hat{\theta}\right)^2 h(\theta_i|r, n).$$

It is straightforward to check that $E[L(\theta, \hat{\theta})|r, n]$ is minimized with respect to $\hat{\theta}$ if we take

$$\hat{\theta} = \sum_{i=1}^{k} \theta_i h(\theta_i|r, n) = E[\theta|r, n],$$

that is, the mean of the posterior distribution. (When it is possible, merely differentiate $E[L(\theta, \hat{\theta})|r, n]$ with respect to $\hat{\theta}$, set the result equal to zero, and solve for $\hat{\theta}$; when the expected loss is not differentiable, the proof is a bit more complicated.)

In a decision-theoretic framework, we use the measure of risk called *admissibility*. [Modern (non-Bayesian) decision theory was formally structured by Abraham Wald; the theory was summarized in Wald, 1950.] An estimator, $\hat{\theta}_A$, of θ is admissible if there is no other estimator $\hat{\theta}$ for which the risk with respect to a loss function $L(\hat{\theta}, \theta|\theta)$ is lower; that is, $\hat{\theta}_A$ is admissible if there is no other estimator $\hat{\theta}$ such that:

(1) $$r(\hat{\theta}|\theta) \equiv EL(\hat{\theta}, \theta|\theta) \leq EL(\hat{\theta}_A, \theta|\theta) \equiv r(\hat{\theta}_A|\theta)$$

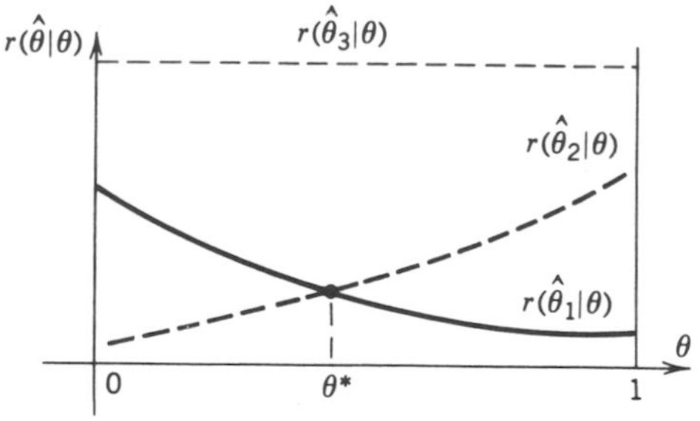

Figure 2.1. Risk functions.

for all θ; and for some $\theta = \theta_0$,

$$r(\hat{\theta}|\theta_0) < r(\hat{\theta}_A|\theta_0). \tag{2}$$

Equivalently, there is no other estimator $\hat{\theta}$ for which

1. $\int L(\hat{\theta}, \theta)p(x_1, \ldots, x_n|\theta)\, dx_1, \ldots, dx_n \leq$
   $\int L(\hat{\theta}_A, \theta)p(x_1, \ldots, x_n|\theta)\, dx_1, \ldots, dx_n$, for all θ; and

2. $\int L(\hat{\theta}, \theta_0)p(x_1, \ldots, x_n|\theta_0)\, dx_1, \ldots, dx_n <$
   $\int L(\hat{\theta}_A, \theta_0)p(x_1, \ldots, x_n|\theta_0)\, dx_1, \ldots, dx_n$, for some $\theta = \theta_0$.

For example, assume we have the risk functions for a continuous θ shown in Figure 2.1. In this example, where $0 \leq \theta \leq 1$, the estimator $\hat{\theta}_3$ is "dominated" by both $\hat{\theta}_1$ and $\hat{\theta}_2$, since its risk is greater than that for either one, for all θ. Moreover, $\hat{\theta}_1$ is not as good as $\hat{\theta}_2$ for $\theta < \theta^*$, whereas it reverses for $\theta > \theta^*$. If no other estimators were possible, both $\hat{\theta}_1$ and $\hat{\theta}_2$ would be admissible, whereas $\hat{\theta}_3$ would be inadmissible.

From a Bayesian point of view it is not relevant to average over the entire sample space $(x_1, \ldots, x_n)$; we are interested in conditioning our inferences only on the current data set (see Section 2.8). Thus, for θ discrete, we choose an estimator by minimizing

$$\sum_{\theta} L(\hat{\theta}, \theta)p(\theta|x_1, \ldots, x_n),$$

that is, the expected loss averaged over the posterior distribution. (This is the result when θ is discrete; in the continuous case we would integrate over the possible values of θ.) The estimator for which this sum (integral) is minimum is the *Bayes' estimator*, and the sum (integral) is the *Bayes' risk*.

Admissibility requires that we average our estimator over all possible values of the observable random variables (the expectation is taken with respect to the observables). In experimental design situations, statisticians must be concerned with the performance of estimators for many possible situational repetitions and for many values of the observables, and then admissibility is a reasonable Bayesian performance criterion. In most other situations, however, statisticians are less concerned with performance of an estimator over many possible samples that have yet to be observed than they are with the performance of their estimator conditional upon having observed this particular data set and conditional upon all prior information available. For this reason, in non-experimental-design situations, admissibility is generally not a compelling criterion for influencing our choice of estimator. We are assured from decision theory, however, that: *as long as the prior distribution is strictly proper, and the Bayesian estimators are unique, they are all admissible*, but not necessarily otherwise (see, e.g., Ferguson, 1967, p. 60). Estimators developed with respect to improper priors (see Section 2.7.2) are not necessarily admissible. Note: distributions are proper if and only if they integrate to unity. Notions of "almost admissible" and "ε-admissible" have been developed to extend the admissibility concept (see for example, Blackwell and Gershick, 1954; or Ferguson, 1967, pp. 63 and 185), but difficulties in three or more dimensions still remain with the use of improper priors (see James and Stein, 1960). Stein (1965) pointed out that (1) all admissible procedures are limits of Bayes' procedures and (2) in many cases, such limits must be formal Bayes' procedures with respect to possibly improper prior distributions.

The judgment formulation (estimation) and decision making procedures discussed above were developed for inferences to be made by individuals. Sometimes, however, groups must make judgments and decisions. For methods of handling such situations, see, for example, Press, 1978, 1980c, 1980d, 1983, 1985a; Press, Ali, and Yang, 1979; Skyrms, 1989; and Hogarth, 1980.

Interval Estimation: Credibility Intervals (Continuous Parameter)

Interval estimators are obtained as probability statements directly from the cdf of the posterior distribution. For example, if we know the posterior cdf of $(\theta | X_1, \ldots, X_N)$ is $F(\theta)$, we can make an interval statement such as:

$$P\{a \le \theta \le b | X_1, \ldots, X_N\} = F(b) - F(a) = .95,$$

so that (a, b) is a 95% "credibility interval" for θ. Note that a and b may depend upon $(X_1, \ldots, X_N)$, but not upon θ.

A credibility interval is the Bayesian version of a confidence interval, except it doesn't have the complicated interpretation generally associated with confidence intervals. Its interpretation, in fact, is the natural one likely to be adopted by most nonexperts using statistics. (See the Note below for interpretation.)

Example: Defective Parts. In a defective parts example (Section 2.6.2) where θ is assumed continuous, the posterior density is given by

$$h(\theta|y) = \frac{1}{B(4,14)}\theta^{4-1}(1-\theta)^{14-1}, \qquad 0 < \theta < 1.$$

Note that $B(\alpha,\beta) = \Gamma(\alpha)\Gamma(\beta)/\Gamma(\alpha+\beta)$ denotes the beta function. Thus, we can make probability statements about θ, conditional on the data, at any desired level of belief. For example, at the .05 level of credibility,

$$\text{if } .05 = P\{0 < \theta < \theta^*|y\}, \quad \text{then } \theta^* = .0846.$$

This result may be obtained from a table of fractiles of the beta distribution. We may obtain a credibility interval for any other levels of credibility in a similar fashion. Moreover, we can also develop two-sided intervals. The interpretation of this credibility statement is that having observed one defective part ($y = 1$) in a sample of size 3 ($n = 3$), and having adopted a beta prior with parameters (3, 12), the probability that θ lies in the interval (0, .0846) is 5%.

We note that the predictive density (Sect. 2.9) can be used in a similar manner to develop credibility intervals for future variables (an example of this is given in Section 5.2.2).

Note: In a confidence interval, we condition on θ, so the interpretation must be that the confidence interval is a random interval, which has meaning only in a long-run probability sense of looking at the fractions of intervals covering the parameter if we took many samples of data of size n not yet observed.

In a credibility interval, we refer directly to the probability of θ being in a preassigned interval, conditional on the data observed in the current experiment. □

We often use the notion of *highest posterior density* (HPD) to determine an appropriate credibility interval. Thus, we seek an interval (a, b) such that:

1. $F(b) - F(a) = .95$.
2. If $p(\theta|X_1,\ldots,X_n)$ denotes the posterior density, for $a \le \theta \le b$, then $p(\theta|X_1,\ldots,X_n)$ is greater than that for any other interval for which condition 1 holds.

Thus, if (a,b) is an HPD interval, for any $\theta_1 \in (a,b)$, and any $\theta_2 \notin (a,b)$, $p(\theta_1|X_1,\ldots, X_n) \geq p(\theta_2|X_1,\ldots, X_n)$, and conversely, subject to condition 1. [This same idea extends to two (and higher) dimensions where we seek HPD regions instead of intervals. For an elaboration of this concept, see Box and Tiao (1965).]

We sometimes view the HPD issue in the opposite way. That is, we sometimes try to find an interval $(a, a + h)$, $h > 0$, such that h is as small as possible, subject to the condition that the cdf $F(\theta)$ obeys the condition

$$F(a + h) - F(a) = 1 - \varepsilon,$$

for some fixed, preassigned ε such as $\varepsilon = .05$. For unimodal distributions it is straightforward to find that if $p(\theta|X_1,\ldots, X_n)$ denotes the density corresponding to $F(\theta)$, then $p(a + h|X_1,\ldots, X_n) = p(a|X_1,\ldots, X_n)$ is the condition for which h is minimum (see Zellner, 1971, p. 27).

For example, suppose

$$p(\theta|X_1,\ldots, X_n) = \frac{1}{B(\alpha, \beta)}\theta^{\alpha-1}(1 - \theta)^{\beta-1},$$

for $\alpha > 0$, $\beta > 0$. For $\alpha = \beta$, we choose HPD credibility intervals by choosing equal tails; this is our simplest case.

For $\alpha \neq \beta$, since $p(\theta|X_1,\ldots, X_n)$ is unimodal, we choose

$$(a + h)^{\alpha-1}(1 - a - h)^{\beta-1} = (a)^{\alpha-1}(1 - a)^{\beta-1},$$

or

$$\left(\frac{a + h}{a}\right)^{\alpha-1}\left(\frac{1 - a - h}{1 - a}\right)^{\beta-1} = 1.$$

This gives a relation for a in terms of h for preassigned α, β. If we also require that

$$F(a + h) - F(a) \equiv \int_a^{a+h} \frac{1}{B(\alpha, \beta)}\theta^{\alpha-1}(1 - \theta)^{\beta-1}\, d\theta = 1 - \varepsilon$$

for, say, $\varepsilon = .05$, we get a second relationship for a in terms of h. Thus, we can use the two relations simultaneously to solve for a and h numerically.

For multimodal distributions the situation is more complicated. We then select the highest mode and solve numerically. The solution may not be unique.

To help us understand our posterior state of knowledge for two-dimensional posterior densities we sometimes plot concentric contours of constant posterior probability or of constant posterior probability density. This is readily accomplished with the aid of a computer, usually by fixing the posterior density at some preassigned value and then plotting the resulting curve. Next choose another posterior density value, and again plot the resulting curve. Repeating the procedure for several values results in a cross-sectional picture of the posterior distribution.

For example, suppose that $\boldsymbol{\theta} = (\theta_1, \theta_2)'$ is bivariate and that $\boldsymbol{\theta} \mid \mathbf{x} \sim N(\mathbf{x}, \mathbf{I})$, where $\mathbf{x} = (2, 3)'$ is observed; that is, $\boldsymbol{\theta}$ given $\mathbf{x}$ is normally distributed with mean vector $\mathbf{x}$ and covariance matrix $\mathbf{I}$. The posterior density is

$$p(\boldsymbol{\theta} \mid \mathbf{x}) = \frac{1}{2\pi} e^{-(\boldsymbol{\theta}-\mathbf{x})'(\boldsymbol{\theta}-\mathbf{x})/2}.$$

Solving for $\boldsymbol{\theta}$ gives

$$(\boldsymbol{\theta} - \mathbf{x})'(\boldsymbol{\theta} - \mathbf{x}) = -2\log[2\pi p(\boldsymbol{\theta} \mid \mathbf{x})],$$

or

$$(\theta_1 - 2)^2 + (\theta_2 - 3)^2 = t^2,$$

where $t^2 \equiv -2\log[2\pi p(\boldsymbol{\theta} \mid \mathbf{x})]$.

For preassigned t, this is the equation of a circle centered at (2, 3). By preassigning a value for the posterior density $p(\boldsymbol{\theta} \mid \mathbf{X})$, or, equivalently, by preassigning t, we will generate a collection of concentric circles for the posterior density contours (see Figure 2.2). They have radius t.

2.3.2 Hypothesis Testing

We are sometimes concerned with the problem of testing a hypothesis H versus an alternative hypothesis A. The classical theory of hypothesis testing has been found by many to be inadequate.

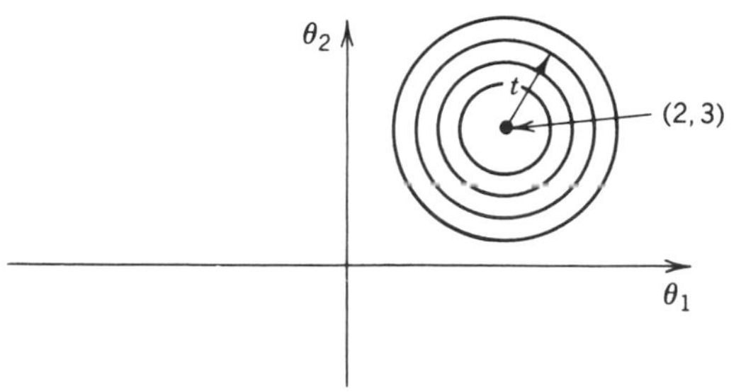

Figure 2.2. Posterior density contours for bivariate normal posterior.

For example, suppose we wish to test H_0: $\theta = \theta_0$ vs. H_1: $\theta \neq \theta_0$, for some known θ_0, and we decide to base our test on a statistic $T \equiv T(X_1, \ldots, X_N)$. It is usually the case that in addition to θ being unknown, we usually can't be certain that $\theta = \theta_0$ precisely, even if it may be close to θ_0. (In fact, in the usual frequentist approach, we often start out believing $\theta \neq \theta_0$, which corresponds to some intervention having had an effect, but we test a null hypothesis H_0 that $\theta = \theta_0$; that is, we start out by disbelieving H_0, and then we test it.) Suppose θ is actually ε away from θ_0, for some $\varepsilon > 0$, and ε is very small. Then, by consistency of the testing procedure, for sufficiently large N we will reject H_0 with probability "one." So, depending upon whether we want to reject H_0, or whether we want to find that we cannot reject H_0, we can choose N accordingly. This is a very unsatisfactory situation. (The same argument applies to all significance testing.) An alternative method of hypothesis testing based upon Bayes' theorem is described below. The method is attributable to Jeffreys (1961, Chapters 5 and 6).

Simple vs. Simple

First we consider the case of testing a simple hypothesis H: $\theta = \theta_0$ against a simple alternative A: $\theta = \theta_1$, where θ_0 and θ_1 are preassigned constants. We assume that H and A are mutually exclusive and exhaustive hypotheses. Let $T \equiv T(X_1, \ldots, X_N)$ denote an appropriate test statistic based upon a sample of N observations. Then, by Bayes' theorem, the posterior probability of H, given the observed data T, is

$$p(H|T) = \frac{p(T|H)p(H)}{p(T|H)p(H) + p(T|A)p(A)},$$

where $p(H)$ and $p(A)$ denote the prior probabilities of H and A. Similarly, for hypothesis A, we have

$$p(A|T) = \frac{p(T|A)p(A)}{p(T|H)p(H) + p(T|A)p(A)}.$$

[*Note:* $p(H|T) + p(A|T) = 1$.] Equivalently,

$$\frac{p(H|T)}{p(A|T)} = \left[\frac{p(H)}{p(A)}\right]\left[\frac{p(T|H)}{p(T|A)}\right].$$

That is, the posterior odds ratio in favor of H is equal to the product of the prior odds ratio and the likelihood ratio. If the posterior odds ratio exceeds unity, we accept H; otherwise, we reject H in favor of A. Note that if there were several possible hypotheses, this approach would extend in a natural way; we would find the hypothesis with the largest posterior probability. We note in passing that the ratio of the posterior odds to the prior odds is called the *Bayes factor*, a factor that depends only upon the sample data. In the case of testing a simple hypothesis vs. a simple alternative, the Bayes factor is just the likelihood ratio.

As an example, suppose $X|\theta \sim N(\theta, 1)$ and we are interested in testing whether H: $\theta = 0$ or A: $\theta = 1$, and these are the only two possibilities. We take a random sample $X_1, \ldots, X_N$ and form the sufficient statistic $T \equiv \bar{X} = (1/N)\Sigma_1^N X_j$. We note that $(T|H) \sim N(0, 1/N)$, and $(T|A) \sim N(1, 1/N)$. Assume that a priori, $p(H) = p(A) = .5$. Then, the posterior odds ratio is given by

$$\frac{p(H|T)}{p(A|T)} = \left(\frac{.5}{.5}\right)\left[\frac{\left(\dfrac{N}{2\pi}\right)^{1/2} e^{-(N/2)\bar{x}^2}}{\left(\dfrac{N}{2\pi}\right)^{1/2} e^{-(N/2)(\bar{x}-1)^2}}\right]$$

$$= e^{-(N/2)[\bar{x}^2 - (\bar{x}-1)^2]}$$

$$= e^{-(N/2)(2\bar{x}-1)}.$$

Suppose our sample is of size $N = 10$, and we find $\bar{x} = 2$. Then, the posterior odds ratio becomes

$$\frac{p(H|T)}{p(A|T)} = e^{-(N/2)(2\bar{x}-1)} = 3.1 \times 10^{-7}.$$

Since the posterior odds ratio is so small, we must clearly reject H in favor of A: $\theta = 1$. Because the prior odds ratio is unity in this case, the posterior odds ratio is equal to the Bayes factor.

Note that comparing the posterior odds ratio with unity is equivalent to choosing the larger of the two posterior probabilities of the hypotheses. If we could assign losses to the two possible incorrect decisions, we would choose the hypothesis with the smaller expected loss.

Simple vs. Composite

Next we consider the more common case of testing a simple hypothesis H vs. a composite hypothesis A. Suppose there is a parameter θ (possibly vector valued) indexing the distribution of the test statistic $T =$

$T(X_1, \ldots, X_N)$. Then, the ratio of the posterior density of H compared with that of A is

$$\frac{p(H|T)}{p(A|T)} = \frac{p(T|H)p(H)}{p(T|A)p(A)} = \frac{p(H)}{p(A)} \frac{P(T|H,\theta)}{\int p(T|A,\theta)p(\theta)\,d\theta},$$

where $p(\theta)$ denotes the prior density for H under A. Thus, the posterior odds ratio, in the case of a composite alternative hypothesis, is the product of the prior odds ratio times the ratio of the averaged likelihoods under H and A. (Note that under H, because it is simple, the liklihood has only one value; so its average is that value.) We assume, of course, that these integrals converge [in the event $p(\theta)$ is an improper density, the integrals will not always exist]. Note also that in this case the Bayes factor is the ratio of the likelihood under H to the averaged likelihood under A.

We note also that we have assumed there are no additional parameters in the problem. If there are, we deal with them by integrating them out with respect to an appropriate prior. For example, suppose $(X|\theta, \sigma^2) \sim N(\theta, \sigma^2)$, H: $\theta = 0$, $\sigma^2 > 0$, vs. A: $\theta \neq 0$, $\sigma^2 > 0$. If $X_1, \ldots, X_n$ are i.i.d., $(\bar{X}, S^2)$ is sufficient for (θ, σ^2), where S^2 is the sample variance. Then the posterior odds ratio for testing H vs. A is:

$$\frac{p(H|\bar{X}, S^2)}{p(A|\bar{X}, S^2)} = \frac{p(H)}{p(A)} \cdot \frac{\int p(\bar{X}|\theta = 0)p(S^2|\sigma^2)p(\sigma^2)\,d\sigma^2}{\int\int p(\bar{X}|\theta)p(\theta)p(S^2|\sigma^2)p(\sigma^2)\,d\theta\,d\sigma^2},$$

for appropriate priors, $p(\theta)$ and $p(\sigma^2)$.

As an example of a simple vs. composite hypothesis testing problem in which there are no additional parameters, suppose $X|\theta \sim N(\theta, 1)$ and we are interested in testing H: $\theta = 0$ vs. A: $\theta \neq 0$. We take a random sample of size 10, $X_1, \ldots, X_N$, $N = 10$, and form the sufficient statistic $T = \bar{X}$ and assume $\bar{X} = 2$. Assume $p(H) = p(A) = .5$. We note that $\bar{X}|\theta \sim N(\theta, 1/N)$, and so

$$p(T|H,\theta) = \left(\frac{N}{2\pi}\right)^{1/2} e^{-(N/2)(\bar{x}^2)}$$

and

$$p(T|A,\theta) = \left(\frac{N}{2\pi}\right)^{1/2} e^{-(N/2)(\bar{x}-\theta)^2}.$$

As a prior distribution for θ under A we take $\theta \sim N(1,1)$. Then,

$$p(\theta) = \frac{1}{(2\pi)^{1/2}} e^{-(\theta-1)^2/2}.$$

The posterior odds ratio becomes

$$\frac{p(H|T)}{p(A|T)} = \frac{\left(\frac{N}{2\pi}\right)^{1/2} e^{-(N/2)\bar{x}^2}}{\int \left(\frac{N}{2\pi}\right)^{1/2} e^{-(N/2)(\bar{x}-\theta)^2} \frac{1}{(2\pi)^{1/2}} e^{-(\theta-1)^2/2}\, d\theta}$$

$$= \frac{(2\pi)^{1/2} e^{-(N/2)\bar{x}^2}}{\int e^{-[(\theta-1)^2 + N(\theta-\bar{x})^2]/2}\, d\theta}$$

$$= (N+1)^{1/2} \exp\left\{\left(-\frac{1}{2}\right)\left[\frac{(N\bar{x}+1)^2}{(N+1)} - 1\right]\right\}.$$

Since $N = 10$, $\bar{x} = 2$, we have

$$\frac{p(H|T)}{p(A|T)} = 1.1 \times 10^{-8}.$$

Thus, we reject H in favor of $A: \theta \neq 0$. That is, the evidence strongly favors A.

For discussion of other hypothesis testing problems, and for discussions of Bayesian hypothesis testing in a decision theoretic framework, see De Groot, 1970 (Chapter 11); Raiffa and Schlaifer, 1961 (Chapter 5); and Schlaifer, 1961 (Chapters 20 and 21).

The Lindley Paradox

The Bayesian approach to testing when little prior information can be developed has been studied (e.g., by Bernardo, 1980; and Lindley, 1965). The topic also received substantial interest when Dennis Lindley called attention to the paradoxical result (Lindley, 1957) that a non-Bayesian could strongly reject a sharp (null) hypothesis H, while a Bayesian could put a lump of prior probability on H and then spread the remaining prior probability out over all other values in a "vague" way (uniformly) and find there are high posterior odds in favor of H. For example, suppose $\bar{y}|\mu \sim N(\mu, 1/n)$, and we wish to test H: $\mu = 0$ vs. A: $\mu \neq 0$. Suppose $P\{\mu = 0\} = P\{\mu \neq 0\}$, and the prior on A is spread out uniformly over $(-M/2, M/2)$. Classically, we would reject H with 95% confidence if $|\sqrt{N}\bar{y}| > 2$. Suppose $n = 40{,}000$, and $\bar{y} = 0.01$ so that $\sqrt{n}\,\bar{y} = 2$, so a sampling theorist would reject H. But the posterior odds ratio is approximately $M(n/2\pi)^{1/2}\exp\{-n\bar{y}^2/2\} = 11$, if $M = 1$, so a Bayesian would accept H. For $\mu = 0$ and $n = 40{,}000$ it is unlikely that $\bar{y} = 0.01$, but it is possible, and such an unlikely event is the reason for the contradictory decision rules. This paradox was taken up again in some detail by Shafer* (1982) and by his discussants.

2.4 BINOMIAL DISTRIBUTION WITH DISCRETE PARAMETER

Let there be n independent trials of an experiment in which there are only two possible outcomes on each trial, say, "success" or "failure." Let X denote the number of successes during the n trials, and let θ denote the probability of success on a single trial. (This was the distribution considered by Bayes; see Appendix 4 for his original essay.) The probability mass function for X is given by

$$f(x|\theta) = \binom{n}{x}\theta^x(1-\theta)^{n-x}, \qquad 0 < \theta < 1, \quad \text{for } x = 0, 1, 2, \ldots, n.$$

2.4.1 Example

Suppose, just to keep matters simple, that θ can have only two possible values, namely, $\theta = .3$ and $\theta = .6$, and suppose that based upon an earlier experiment you believe that

$$P\{\theta = .3\} = .1 \quad \text{and} \quad P\{\theta = .6\} = .9.$$

So the prior pmf is

$$g(\theta) = \begin{cases} .1, & \theta = .3 \\ .9, & \theta = .6 \end{cases}.$$

By Bayes' theorem, the posterior probability mass function is given by

$$h(\theta|x) \propto \theta^x(1-\theta)^{n-x}g(\theta) \quad \text{for } \theta = .3, .6,$$

*Shafer advocates the use of (1) the "belief functions" (see Shafer, 1976) based upon the notion of nonadditive degrees of belief and (2) the "rule of combination" developed by Dempster (1967), using "lower and upper probabilities" (see also Section 2.7.1).

or

$$h(\theta|x) = \frac{\theta^x(1-\theta)^{n-x}g(\theta)}{(.3)^x(.7)^{n-x}g(.3) + (.6)^x(.4)^{n-x}g(.6)}, \qquad \theta = .3, .6.$$

Suppose $n = 5$ and $x = 2$. Then we can find, for example,

$$h(.3|x) = \frac{(.3)^2(.7)^{5-2}g(.3)}{(.3)^2(.7)^{5-2}g(.3) + (.6)^2(.4)^{5-2}g(.6)} = .1296 \cong .13.$$

Similarly,

$$h(.6|x) = .8704 \cong .87.$$

Thus, the posterior probability mass function for θ is

$$h(\theta|x) = \begin{cases} .13, & \theta = .3 \\ .87, & \theta = .6 \end{cases}.$$

Because there were only five data points, our posterior pmf does not differ very much from our prior pmf; that is, we need more evidence to the contrary before we'll greatly alter our pre-experiment beliefs that were based upon earlier practical experience.

The idea of assigning a loss of K to an incorrect decision, and a loss of zero to a correct decision, and then adopting a decision procedure which minimizes expected loss, can readily be extended to problems of testing simple vs. composite hypotheses.

2.4.2 Solution for Defective Parts Example

We now return to the problem involving eight parts shipped to a supplier (Section 2.2.1), for which we asked the question "what are the chances the box contains two defective parts when a sample of three contains one defective part?" First note that the number of defective parts X in a sample of size n of parts in a box approximately follows a binomial distribution (actually, a hypergeometric distribution but we will ignore this correction), so

$$P\{X = x|\theta, n\} \equiv f(x|\theta) = \binom{n}{x}\theta^x(1-\theta)^{n-x},$$

where θ denotes the probability of a defective part.

Here, for one defective part to be present, for example, when $n = 3$, the likelihood function is

$$f(1|\theta) = \binom{3}{1}(\theta)(1-\theta)^2 = 3\theta(1-\theta)^2.$$

There are three possibilities: There are 0, 1, or 2 defective parts in a box of 8; or since θ denotes the probability of a defective part, we have $\theta = 0, 1/8, 2/8$. Our prior belief, based upon past experience, is

Defective probability: θ	0	.125	.25
Probability mass function: $p(\theta)$	.70	.20	.10

Moreover, for these three values of θ, the likelihood function is given by

θ	0	.125	.25
$f(1\|\theta)$	0	.287	.42

Bayes' theorem gives:

$$\begin{aligned} &P\{\text{Box contains 2 defective parts}|\text{sample contains 1 defective}\} \\ &= P\{\theta = .25|X = 1\} \\ &= \frac{f(1|.25)p(.25)}{f(1|.25)p(.25) + f(1|.125)p(.125) + f(1|0)p(0)} \\ &= \frac{(.42)(.10)}{(.42)(.10) + (.287)(.20) + (0)(.70)} = .424 \\ &= 42\%. \end{aligned}$$

In this problem, while the prior probability of two defectives in the shipment is only 10%, the posterior probability is four times as much (42%). (See also Section 2.2.1.)

2.5 BAYES' THEOREM (CONTINUOUS PARAMETER)

Let $\mathbf{X}_1, \ldots, \mathbf{X}_N$ denote a sample from a distribution with likelihood function L indexed by a continuous parameter $\boldsymbol{\theta}$, with prior density $g(\boldsymbol{\theta})$. Then, the posterior density for $\boldsymbol{\theta}$ is given by

$$h(\boldsymbol{\theta}|\mathbf{x}_1, \ldots, \mathbf{x}_N) = \frac{L(\mathbf{x}_1, \ldots, \mathbf{x}_N|\boldsymbol{\theta})g(\boldsymbol{\theta})}{\int L(\mathbf{x}_1, \ldots, \mathbf{x}_N|\boldsymbol{\theta})g(\boldsymbol{\theta})\, d\boldsymbol{\theta}}.$$

Moreover, since the denominator does not depend upon $\boldsymbol{\theta}$, we have

$$h(\boldsymbol{\theta}|\mathbf{x}_1,\ldots,\mathbf{x}_N) \propto L(\mathbf{x}_1,\ldots,\mathbf{x}_N|\boldsymbol{\theta})g(\boldsymbol{\theta}).$$

Proof: By a conditional probability argument, as in the discrete variable case. □

2.6 EXAMPLE: BINOMIAL DISTRIBUTION WITH CONTINUOUS PARAMETER

$$f(y|\theta) = \binom{n}{y}\theta^y(1-\theta)^{n-y}, \qquad 0 < \theta < 1, \quad y = 0,1,\ldots,n.$$

Since $\binom{n}{y}$ does not depend upon θ, if we take $L(x_1,\ldots,x_N|\theta) \equiv f(y|\theta)$, where the x_i's denote Bernoulli variables (with values equal to 0 or 1), and $y = \Sigma_{i=1}^n x_i$, Bayes' theorem gives the posterior density

$$h(\theta|y) = \frac{\theta^y(1-\theta)^{n-y}g(\theta)}{\int_0^1 \theta^y(1-\theta)^{n-y}g(\theta)\,d\theta}.$$

2.6.1 Uniform Prior

Suppose

$$g(\theta) = \begin{cases} 1, & 0 < \theta < 1 \\ 0, & \text{otherwise} \end{cases},$$

then

$$h(\theta|y) = \frac{1}{B(y+1, n-y+1)}\theta^y(1-\theta)^{n-y};$$

that is, the posterior distribution of θ given y is beta (centered at the modal value $\theta = y/n$), and the mean of θ given y is $E(\theta|y) = (y+1)/(n+2)$.

2.6.2 Defective Parts Example

Here, θ denotes the probability of a defective part. Moreover, if y denotes the number of defective parts in the sample, the likelihood is

$$L(y|\theta) = \theta^y(1-\theta)^{n-y}, \qquad 0 < \theta < 1,$$

and the posterior density is

$$h(\theta|y) = \frac{\theta^y(1-\theta)^{n-y}g(\theta)}{\int_0^1 \theta^y(1-\theta)^{n-y}g(\theta)\,d\theta},$$

where $g(\theta)$ denotes the prior density.

Suppose we use our past experience to conclude a priori that

$$E(\theta) = .2, \qquad \sigma_\theta^2 = \text{var}(\theta) = .01 \qquad (\sigma_\theta = .1).$$

Then, if we are comfortable permitting our beliefs about the unknown θ to follow a beta distribution,

$$g(\theta) = \frac{1}{B(\alpha,\beta)}\theta^{\alpha-1}(1-\theta)^{\beta-1}, \qquad 0<\theta<1, \quad 0<\alpha, \quad 0<\beta,$$

we can solve backwards for (α, β) from the two equations:

$$E(\theta) = .2 = \frac{\alpha}{\alpha+\beta}, \qquad \text{var}(\theta) = .01 = \frac{\alpha\beta}{(\alpha+\beta)^2(\alpha+\beta+1)},$$

and find that $\alpha = 3$, $\beta = 12$. The mode of the prior density is at $(\alpha - 1)/(\alpha + \beta - 2) = 2/13 \cong .15$.

Thus, the prior density looks like Figure 2.3. The posterior density becomes

$$h(\theta|y) = \frac{1}{B(y+\alpha, n-y+\beta)}\theta^{y+\alpha-1}(1-\theta)^{n-y+\beta-1},$$

which, for $y = 1$ (number of defectives in the sample) and $n = 3$ (sample size), looks like (see Figure 2.4)

$$h(\theta|y) = \frac{1}{B(4,14)}\theta^3(1-\theta)^{13}, \qquad 0<\theta<1,$$

so that the data have not changed our prior belief very much. (Compare Figure 2.3 with Figure 2.4.)

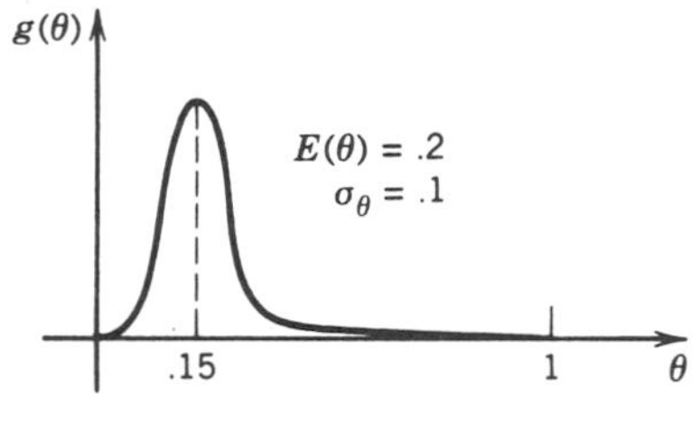

Figure 2.3. Prior density for θ (probability of a defective).

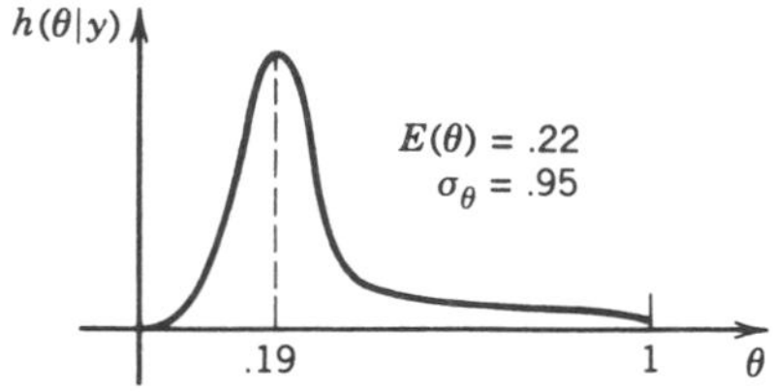

Figure 2.4. Posterior density for θ.

2.7 PRIOR DISTRIBUTIONS

2.7.1* The Bayesian vs. the Non-Bayesian Approach to Inference: Empirical Bayes Estimation

Mainly because of its use of subjective prior beliefs, the approach to statistical inference based upon Bayes' theorem has been controversial within the field of statistics for many years (a detailed discussion will follow later in this section). The controversy has engendered a schism of foundational beliefs concerning the most appropriate basis for developing statistical procedures for solving problems of inference. One school of thought holds that the Bayesian approach is the only simultaneously axiomatic (recall Savage's axiom system discussed in the Complement to Chapter I), self-consistent, and intuitively sensible system we have. If everyone believes this assertion, then why isn't everyone a Bayesian? [Efron (1986) proposed a partial "answer" by claiming that inferences should be objective, not subjective; other answers are offered below, in this section.]

Within the Bayesian framework there is a wide spectrum of beliefs. At one end of the spectrum is orthodoxy (strict adherence to the assumptions and implications of Bayes' theorem).

In the middle of the Bayesian spectrum is James–Stein (1960) estimation, and "empirical Bayes" inference (see, e.g., Robbins, 1955†; Maritz, 1970; Efron and Morris, 1973 and 1977; and Berger, 1980). In the most frequently adopted version of empirical Bayes estimation the Bayesian estimation structure is used with a preassigned family of prior distributions to obtain the Bayes estimator. But the parameters of the prior distribution,

*Section 2.7.1 is philosophical in nature and may be skipped in a first reading, especially by readers desiring a more "methods-oriented" approach to the subject.

†The term "empirical Bayes" was coined by Herbert Robbins (1955), who used the term in the context in which the same type of estimator or decision was required repeatedly in a large sequence of problems.

also called *hyperparameters*, are not assessed subjectively; rather, they are estimated from the current data set by using the marginal distribution of the data, given the hyperparameters. Often, the hyperparameters are estimated by maximum likelihood or by sample moments. The data-based estimators of the hyperparameters are then substituted for the hyperparameters in the prior distribution, as well as in the Bayesian estimator, to obtain an *approximate Bayesian estimator*, also referred to as an *empirical Bayes estimator*. This approach often has the virtue of providing estimators that may be shown to be admissible with respect to quadratic (or other reasonable) loss functions. Moreover, the empirical Bayes approach yields estimators that bound the risk when using quadratic (or other unbounded) loss functions. Unfortunately, however, the empirical Bayes procedure violates Bayes' theorem and therefore, the laws of probability, which require that the prior distribution not depend upon the current data set. That is, the assumptions of conditional probability, upon which Bayes' theorem is founded, require that the prior distribution depend only upon its parameters and not upon the data; otherwise the theorem doesn't hold. Thus, even though it is incoherent to use these estimators (see Section 1.3.2), empirical Bayes estimators can be viewed as approximate Bayes estimators. One problem with these estimators is that because there is no natural standard error, it is difficult to find credibility intervals or test hypotheses.

As an illustration, suppose $\mathbf{x} \mid \boldsymbol{\theta} \sim N(\boldsymbol{\theta}, \mathbf{I})$, and as a prior we take $\boldsymbol{\theta} \sim N(\boldsymbol{\mu}, \tau^2 \mathbf{I}_k)$. The posterior becomes $\boldsymbol{\theta} \mid \mathbf{x} \sim N(\boldsymbol{\theta}_B, \phi^2 \mathbf{I})$, where:

$$\boldsymbol{\theta}_B = \frac{\mathbf{x}}{1+\tau^{-2}} + \left(\frac{\tau^{-2}}{1+\tau^{-2}}\right)\boldsymbol{\mu}, \qquad \phi^2 = \frac{\tau^2}{1+\tau^2}.$$

The Orthodox Bayesian assesses $(\boldsymbol{\mu}, \tau^2)$; the empirical Bayesian estimates τ^2 from the sample data, and guesses $\boldsymbol{\mu}$. Often, he takes $\mu_i = 1/k\Sigma_1^k x_i$, $\boldsymbol{\mu} = (\mu_i)$, $\mathbf{x} = (x_i)$. To estimate τ^2, note that unconditionally, $\mathbf{x} \sim N(\boldsymbol{\mu}, (1+\tau^2)\mathbf{I})$. Then, if $S \equiv (\mathbf{x} - \boldsymbol{\mu})'(\mathbf{x} - \boldsymbol{\mu})$, $S \sim (1+\tau^2)\chi_k^2$, and S^{-1} follows an inverted gamma distribution. So $E[(k-2)/S] = (1+\tau^2)^{-1}$. This gives $(1+\tau^2)^{-1} = (k-2)/S$. This yields the empirical Bayes estimator, $\hat{\theta}_B$, with components:

$$\hat{\theta}_{B_i} = \mu_i + \left(1 - \frac{k-2}{S}\right)(x_i - \mu_i); \qquad i = 1, \ldots, k.$$

Often, the risk of the empirical Bayes estimator is less than that of the

MLE, but it is always greater than that of the Bayes estimator. Note that the components of **x** must be in the same units, and for a good estimator, the θ_i should be close to one another.

At the other end of the philosophically Bayesian spectrum are various schools of thought attempting to generalize the entire Bayesian framework to provide alternative procedures for updating degrees of belief to handle situations that they feel are not handled well within the strictly Bayesian framework, such as updating a belief on the basis of new information that is "probable," or uncertain, as opposed to new data that is observed.

For example, suppose we are trying to assess the probability of the home team winning in tomorrow's game, and we are told that their most valuable player "may not" play because of a sudden injury. For situations like this, Dempster (1967) developed a new probability axiom system based upon "lower and upper probability" (using multiple-valued transformations for event sets) with a nonadditive rule for combining probabilities; and Shafer (1976), using the Dempster model, developed a new probability system of "belief" functions. This approach, while difficult to implement, generalizes the Bayesian framework.

The classical (frequentist, sampling theory) approach to statistical inference is based upon the concepts and principles developed by the founders of the field in the early 20th century, namely, R. A. Fisher, Jerzy Neyman, and Karl Pearson. There are other systems of inference and other corresponding schools of thought that have developed, such as fiducial inference (R. A. Fisher) and structural inference (D. S. Fraser). Those who are interested in pursuing the comparisons of these various systems should see, for example, Barnett (1982).

Schisms about foundational beliefs are common in science, as is well known, for example, to those who have followed the debate that has raged within physics for over half a century. This debate is over the classical deterministic view of the nature of the universe (as it was seen, for example, by Albert Einstein), as opposed to the view promulgated by Max Born, Werner Heisenberg, and Paul Dirac, that the universe is inherently indeterminate in description and can best be interpreted statistically and quantum mechanically. Does light consist of particles or waves? Are electrons particles or waves? A dualism of interpretation has developed so that now the two positions appear to coexist somehow.

In statistics, the Bayesian approach has enjoyed a renaissance during the second half of the 20th century, having been viewed with quiet, uninterested tolerance for the 165 years that elapsed between Laplace (in 1774) and Jeffreys (in 1939). (There were other interested researchers in between Laplace and Jeffreys, but none appears to have been as vocal or cogent in his scientific presentations.) But the methodological developments in the field have enjoyed tremendous growth during the last few decades, and the number of scientists who not only accept the approach but who also argue for, and contribute to, its continued development has steadily increased.

The reasons for the schism within the field of statistics over the Bayesian position are discussed below.

1. *Objectivism vs. Subjectivism.* Many statistical (and other) scientists feel that science should be as objective as possible, and personal feelings and beliefs should not enter into scientific calculations. For this reason alone, such people are inclined to reject a methodology purporting to be scientific, which is based upon subjectivism.

In fact, most scientific inference is already subjective, and it always has been. When a scientist decides to do a particular experiment, it is often because he/she has preconceived notions about what he/she is likely to find. In the formalism of the scientific method, this preconceived notion is called the *hypothesis*. The experiment is called a *confirmatory experiment* in that the scientist is trying to confirm (or deny) a particular hypothesis. (Some "experiments" are not confirmatory but are *exploratory*, in that the experimenter not only doesn't know what he/she will find but doesn't even have a hypothesis about what he/she will find.)

Once a confirmatory experiment is carried out, some data points are sometimes excluded from analysis for "subjective reasons" (they are thought to be "mistakes"; they are thought to be too large or too small to be of interest; or experimental conditions were somehow different for those particular observations). In fact, oftentimes such data should be included in the analysis, even if it means a simple theory will not fit the total data set.

If the hypothesis is to be tested, the experimenter must decide subjectively upon a level of significance for rejection of the hypothesis. Should he/she use a 5% level, which, in a specific case, might imply he/she should reject the hypothesis, whereas if he/she were to use a 1% level of significance he/she would conclude he/she could not reject the hypothesis. If the experimenter strongly believes in the hypothesis, a priori, he/she may sometimes be inclined to use the significance level that supports his/her hypothesis. Using p-values will not solve the problem. In fact, p-values can be misleading, in that we are often led to believe that a result is significant, say, rejecting H_0, at some level of significance, say, $p = .05$, when a Bayesian test of the hypothesis would indicate that the sample evidence favors H_0. Even when a Bayesian test would also not support H_0, the posterior evidence against H_0 is generally weaker than that which would be reflected by the p-value. For a more elaborate explanation of these ideas, see Berger and Sellke, 1987; also see Casella and Berger, 1987; and Edwards, Lindman, and Savage, 1963. The experimenter will probably cling to his/her preconceived belief until other experiments negate the conclusions.

The decision about which experiment the researcher should choose to carry out is subjective, as is the model he/she elects to use to analyze the data.

In the final analysis it's a matter of sample size. In large samples, the prior plays little role and the data "speak" most loudly. In this instance, prior belief is not involved and subjectivity disappears. It is only in small samples where there is insufficient data to form a strong conclusion that the prior beliefs of those most competent to judge become important.

In summary, subjectivism is already well instituted in science, and rightly so, because the experimenter knows more about that particular experiment than anyone else, and the public is generally willing to accept the experimenter as the expert. They must continue to demand, however, that the experiment be repeated by others and that the experimenter's results be subjected to peer review.

2. *Choice of the Public Policy Prior.* In many situations the result of a scientific experiment must be reported to the public as a general result, and public policy may be based upon this finding. For example, we may wish to know what percentage change there will be in the number of "high-quality" young people who will enlist in a volunteer army if they are offered a $5000 bonus for enlisting for at least 4 years. A particular individual may have a prior belief about this percentage, and an experiment may be carried out to study the problem, but in the final analysis, the army does not want posterior inferences made by different people to yield different answers (unless the different assumptions they use to develop their posteriors are clearly stated). This uniform inference requirement is quite reasonable for many problems. What should the statistician do in such situations when he/she wants to choose a "public policy prior" (one which all observers will accept)? Some classical statisticians have been leery of adopting the Bayesian approach in such situations because of their concern about which prior to choose. As will be seen below in such situations, the Bayesian statistician either will use a "vague prior" or will report a set of priors and their corresponding posteriors. That is, he/she will report a mapping from the priors to the implied posteriors.

In some public policy situations we don't necessarily want a common prior; we merely want to know the policy implications of particular prior beliefs. Dickey (1973) proposed presenting the posterior as a function of the prior to cover such situations. Although a single prior is not selected, the public is provided with the posterior inferences associated with a range of priors.

Pierre Simon Laplace (1749–1827) was also concerned with the public policy prior. He stated what we now call Bayes' theorem (Laplace, 1774, p. 623), 11 years after Bayes' essay appeared, in the following form (this translation is from Stigler, 1986):

> If an event can be produced by a number n of different causes, then the probabilities of these causes given the event are to each other as the probabilities of the event given the causes, and the probability of the existence of each of these is equal to the probability of the event given the cause, divided by the sum of all the probabilities of the event given each of these causes.

Here, Laplace is taking the a priori probabilities to be equal, which was his early approach to dealing with prior beliefs (he later generalized the prior). (Recall from Section 1.4 that Laplace is not likely to have seen Bayes' essay, which appeared originally in England in 1763 and probably did not circulate in France until about 1780. Thus, Laplace appears to have solved the problem of inverse probability independently of Bayes, but 11 years later.)

Laplace (1812) formally proposed that in situations where we need to use a public policy prior (he actually used words more appropriate to his time), and where we would therefore like to proceed by introducing as little external subjective information into the problem as possible (beyond the sample data obtained from an experiment), we should use the *principle of insufficient reason*. His principle suggested that in the absence of any reason to the contrary, all values of the unknown parameter indexing the occurrence of events should be taken to be equally likely, a priori. (In the second edition of his book, in 1814, he modified the principle to include unequal prior probabilities, if there were sufficient reason.)

Jeffreys (1961) (and in the earlier editions of his book, in 1939 and 1948) followed essentially the same insufficient reason principle, but he modified it slightly for parameters, with support only on the positive half of the real line (these ideas will be given mathematical form in Section 2.7.2). He also extended the principle to multiparameter families (see also Press, 1982), and he used invariance arguments to justify the procedures. Jaynes (1983) and Zellner (1971) recommended the same Laplace (or modified Laplace) procedure for choosing a public policy prior distribution. This prior can be thought of as "objective," in that all observers will adopt the same prior as representing a prior state of indifference to one parameter value over another. Moreover, very often, use of such a prior will yield statistical procedures analogous to those that are likely to be used by non-Bayesian statisticians, procedures developed using classical (frequentist) principles.

When the unknown parameter lies on a finite interval, the uniform distribution serves us well as the public policy, or objective, prior. When at least one end point of the domain of the parameter is not finite, however, the objective prior implied becomes improper. The fact that the prior distribution is improper, even though the posterior distribution may be perfectly proper, has led to some skepticism about this part of the Bayesian approach (we need to focus on the Rényi probability axiom system to legitimize these improper priors in those situations where they are appropriate; see last paragraph of Section 1.3.1).

3. *Lack of Sufficient Evidence.* Some statisticians feel that the evidence against the frequentist approach to statistical inference and decision making is not yet sufficiently cogent to warrant discarding a large collection of procedures that by and large have worked pretty well. While it is true that many frequentist procedures that have been developed have yielded very reasonable results in many contexts, there are also many instances of frequentist procedures yielding unreasonable results.

As a simple example of inconsistencies in the frequentist approach, consider the problem of estimating θ^2 from independent identically distributed (i.i.d.) observations $X_1, \ldots, X_n$ from $N(\theta, 1)$. The sampling theory estimator on which attention will focus most heavily is $\hat{\theta}^2 \equiv (\bar{X}^2 - 1/n)$ (where $\bar{X}$ denotes the sample mean), because it is the uniformly minimum variance unbiased estimator. Of course for small n, $\hat{\theta}^2$ could be negative while we know we are estimating a positive quantity. Other interesting examples of strange results and procedures implied by the frequentist approach may be found in Basu (1964, 1988), and in Robinson (1975).

The public policy, or objective, prior distribution discussed above is given mathematical form as the "vague" prior in Section 2.7.2.

2.7.2 Vague (Indifference) Priors

You can develop a prior distribution representing your degree of belief for any quantity unknown to you.

Suppose you are asked to assess a prior distribution for the density (weight per unit volume) of potassium in your bloodstream, θ. Unless you have a strong background in biology, at first you may throw up your hands and say "I am totally uninformed about θ," and then you may take all values of θ as equally likely. But when you think about it a bit you realize that there are really some things you know about θ.

For example, since you know that a potassium density can't be negative, you know $\theta > 0$. You might take the prior density

$$g(\theta) = \begin{cases} \frac{1}{a}, & 0 < \theta < a, \\ 0, & \text{otherwise}. \end{cases}$$

But what should you take for a? And what would this imply about your prior knowledge of $[\log \theta]$, and $[\theta^2]$? (It will be seen later in this section that for $\theta > 0$ we will take $\log \theta$ to be uniform, which implies that $g(\theta) \propto \theta^{-1}$; this represents vagueness about θ.)

Suppose, for illustration, we consider the problem in which $X|\theta \sim N(\theta, 1)$, $-\infty < \theta < \infty$, and we wish to estimate θ. A feeling of being totally uninformed about the true value of θ might lead some of us to an improper prior distribution of θ over the entire real line. If we demand the use of only proper priors, we are led to confront an impossible situation, as we see below. It will be seen that if we are "uninformed" about θ, we necessarily become "informed" about some functions of θ, and conversely.

Suppose we try to argue that to be indifferent to all values of θ should imply that if we transform θ by projecting it in a simple, monotonically smooth way into the unit interval and then put a uniform (proper) prior distribution on all values in the unit interval, we will have circumvented our problem. Not so, as we shall see below.

Our problem is to develop a prior that will reflect indifference to all values of θ.

Let

$$\phi \equiv F(\theta),$$

where $F(\theta)$ denotes any monotone, nondecreasing, differentiable transformation of θ such that

$$0 \leq F(\theta) \leq 1.$$

For example, $F(\cdot)$ can be any cdf. Let the prior distribution for ϕ be uniform; that is, its density is

$$p(\phi) = 1, \qquad 0 < \phi < 1.$$

Clearly, $p(\phi)$ reflects indifference to all values of ϕ. But the inverse transformation $\theta = F^{-1}(\phi)$ induces a prior density on θ,

$$g(\theta) = \left|\frac{dF(\theta)}{d\theta}\right|,$$

by the basic theorem on transformation of densities. We are free to choose any $F(\cdot)$. Suppose we choose the normal cdf:

$$\phi \equiv F(\theta) = \int_{-\infty}^{\theta} \frac{1}{\sqrt{2\pi}} e^{-t^2/2}\, dt.$$

Then, the density of θ is

$$g(\theta) = \frac{1}{\sqrt{2\pi}} e^{-\theta^2/2},$$

or $\theta \sim N(0, 1)$. Then the posterior density is

$$p(\theta|X) \propto e^{-[(x-\theta)^2/2]-(\theta^2/2)}$$

or

$$\theta|x \sim N\left(\frac{x}{2}, \frac{1}{2}\right).$$

But note that we are being extremely informative about θ, in spite of assigning equal probability to all possible intervals of ϕ of equal length; and all we did was try to be indifferent (Laplacian) to all values of a simple monotone transformation of θ. In order for the prior distribution of θ to express indifference toward specific values of θ, we would need to have

$$g(\theta) \propto \text{constant},$$

which, in turn, implies that $dF(\theta)/d\theta$ is constant, or $F(\theta)$ is a linear function of θ. But it is impossible (in the usual probability calculus) for θ on the real line (or even on the positive real line) to have $0 \leq F(\theta) \leq 1$, and have $F(\theta)$ be a straight line for $-\infty < \theta < \infty$.

Placing uniform distributions on simple monotone transformations of θ will not solve our problem, nor will more complicated transformation procedures. The problem is more basic, and it exhorts us to develop some sort of probability algebra in which it is reasonable to adopt improper prior distributions on θ over the entire real line (see last paragraph of Section

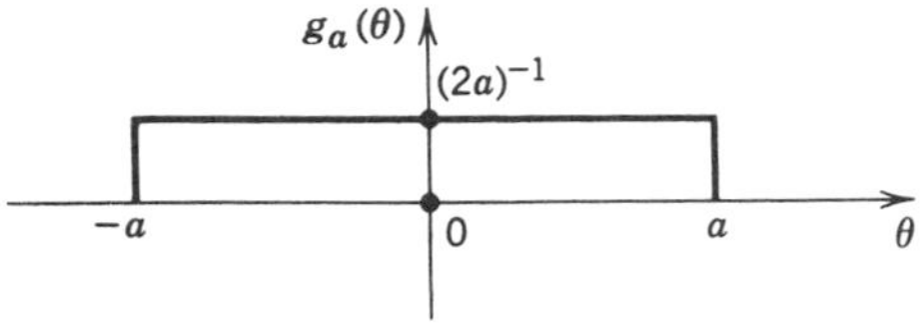

Figure 2.5. Vague prior density on $(-a, a)$.

1.3.1). In the sequel, whenever we want to express the belief that, a priori, no value of θ is any more likely than any other value, or that you are indifferent to one value over any other, we will adopt a "vague prior," which is explained below.

To express vagueness, or indifference about our knowledge of θ, when θ is a parameter lying on the real line $(-\infty < \theta < \infty)$, we use an improper distribution, called *vague*, of the form

$$g(\theta) \propto \text{constant},$$

which is a limiting form of $g_a(\theta)$, where

$$g(\theta) = \lim_{a \to \infty} g_a(\theta)$$

and (see Figure 2.5) where

$$g_a(\theta) = \begin{cases} \dfrac{1}{2a}, & -a < \theta < a, \\ 0, & \text{otherwise.} \end{cases}$$

(*Note:* The limiting process used to arrive at a vague prior is not unique.)

Principle of Stable Estimation

Savage, 1962, noted that by Bayes' theorem, the posterior density depends only upon the *product* of the likelihood function and the prior. Therefore, he suggested that to be vague about θ we need only take the prior on θ to be uniform over the range where the likelihood function is non-negligible. Outside of this range, we can take the prior to be any convenient smooth function. It won't matter which function because the posterior density will be close to zero regardless.

For a positive parameter, say σ, we take $\log \sigma$ to be uniformly distributed on the entire real line, as we did previously with θ. By transforming variables we find that (using the same g, generically, to mean "density function," but not the same one that was used for θ), the prior density becomes

$$g(\sigma) \propto \frac{1}{\sigma}, \text{ and } g(\sigma^2) \propto \frac{1}{\sigma^2}.$$

These vague priors have properties of invariance of inferences under various simple group structures. That is, if the parameter θ or σ is transformed, posterior inferences based upon the new parameter will be consistent with those based upon the old parameter. Thus, if $g(\theta)$ is the prior density for θ, and if we transform θ to ϕ by $\phi = F(\theta)$ and if $h(\phi)$ is the prior density for ϕ, we should have $g(\theta)\, d\theta = h(\phi)\, d\phi$, and posterior inferences in either case should reflect this consistency. There have been various proposals for how to express the notion of indifference to one parameter value over another by using group invariance (see, e.g., Hartigan, 1964; Jeffreys, 1961, pp. 179–181; Villegas, 1969). Other principles for selecting prior distributions that reflect indifference have been based upon information theory (see Bernardo, 1979; Jaynes, 1983, pp. 116–130 "Prior Probabilities," plus other related articles; and Zellner, 1971, pp. 50–51). The reason for the diversity of views on the subject is that the method for expressing the notion of "knowing little" differs depending upon the context. For this reason, various authors propose different priors for different groups of transformations appropriate in different settings.

2.7.3 Natural Conjugate Priors; *g*-Priors

Suppose we are not vague about θ but have some (non-equally likely) prior beliefs. It will often suffice to permit our prior beliefs to be represented by some smooth distribution that is a specific member of a family of prior distributions which have convenient mathematical properties. One such family is called a *natural conjugate family*. (Such a prior is also sometimes called a *convenience prior*.)

Example

Suppose the likelihood function is binomial, so that

$$L(\theta) = \theta^y (1-\theta)^{n-y}.$$

(Recall that the likelihood function is unique only up to a multiplicative constant.) We develop the family of natural conjugate priors by interchanging the roles of the random variable and the parameter in the likelihood function and then "enriching" the parameters, that is, making their values perfectly general and not dependent upon the current data set. Thus, if θ were the random variable, $L(\theta)$ would look like a *beta distribution*. But we don't want the beta prior family to depend upon sample data (y and n), so we use arbitrary parameters (α, β), and we norm the density to make it proper, to get the beta prior density

$$g(\theta) = \frac{1}{B(\alpha, \beta)}\theta^{\alpha-1}(1-\theta)^{\beta-1}, \qquad \alpha > 0, \quad \beta > 0.$$

Now we use our prior beliefs to assess the hyperparameters α and β; that is, having fixed the family of priors as the beta distribution, only (α, β) remain unknown (we do not have to assess the entire prior distribution).

An additional mathematical convenience arises in computing the posterior. Since the posterior density is proportional to the likelihood times the prior, in this case a beta prior, the posterior density is given by

$$h(\theta|y) \propto \left[\theta^{y}(1-\theta)^{n-y}\right]\left[\theta^{\alpha-1}(1-\theta)^{\beta-1}\right],$$

$$h(\theta|y) \propto \theta^{y+\alpha-1}(1-\theta)^{n-y+\beta-1},$$

a beta posterior; many natural conjugate prior distributions have the advantage that the posterior distributions have the same form (distributional family) as the prior distributions. This "closure property" for some distributions of the natural conjugate family holds for all sampling distribution members of the exponential family, that is, the family of distributions with sufficient statistics (subject to some simple regularity conditions).

Table 2.1 provides sampling distributions in the exponential family, along with their corresponding natural conjugate priors. The reader should try to establish these relations by following the example of the binomial, above.

Data Based Priors

If $x_1, \ldots, x_n$ are i.i.d. from the exponential distribution with density $f(x|\theta) = \theta\exp(-\theta x)$, and if we adopt the vague prior $p(\theta) \propto \theta^{-1}$, the posterior density becomes

$$h(\theta|x_1, \ldots, x_n) \propto \theta^{n-1}\exp(-n\bar{x}\theta).$$

If $y_1, \ldots, y_m$ is a new data set from $f(x|\theta)$, and we use $h(\cdot)$ as a prior since it is natural conjugate with hyperparameters based upon the earlier data set (it is a data based prior), the posterior becomes

$$p(\theta|y_1, \ldots, y_m; x_1, \ldots, x_n) \propto \theta^{m+n-1}\exp\{-\theta(m\bar{y} + n\bar{x})\},$$

Table 2.1 Some Natural Conjugate Prior Distributions

Sampling Distribution	Natural Conjugate Prior Distribution
1. Binomial	Success probability is beta
2. Negative binomial	Success probability is beta
3. Poisson	Mean is gamma
4. Exponential with mean λ^{-1}	λ is gamma
5. Normal with known variance but unknown mean	Mean is normal
6. Normal with known mean but unknown variance	Variance is inverted gamma

a density in the same family (gamma), but now based upon an equivalent set of $m + n$ observations. So our experience accumulates as additional data.

***g*-Priors**

A variant of the natural conjugate prior family that is sometimes used in regression problems is the "g-prior" distribution family (see Zellner, 1980). This family takes into account the form of the experiment used to generate the observational data.

Suppose

$$(\mathbf{y} \mid \mathbf{X}) = \mathbf{X}\boldsymbol{\beta} + \mathbf{u}$$

denotes a univariate regression, where $\mathbf{y}: (n \times 1)$ denotes the vector of dependent variable observations, $\boldsymbol{\beta}$ denotes an unknown regression coefficient vector, $\mathbf{u}$ denotes an n-vector of disturbance terms, with $E(\mathbf{u}) = \mathbf{0}$, $\text{var}(\mathbf{u}) = \sigma^2\mathbf{I}$, and $\mathbf{X}$ denotes a matrix of explanatory (independent) variables fixed by the design of the experiment generating the $\mathbf{y}$ vector. We adopt the g-prior distribution family

$$p(\boldsymbol{\beta}, \sigma) = p_1(\boldsymbol{\beta} \mid \sigma) p_2(\sigma),$$

where

$$p_1(\boldsymbol{\beta} \mid \sigma) \propto \frac{1}{\sigma^k} \exp\left\{\left(-\frac{g}{2\sigma^2}\right)(\boldsymbol{\beta} - \bar{\boldsymbol{\beta}})'(\mathbf{X}'\mathbf{X})(\boldsymbol{\beta} - \bar{\boldsymbol{\beta}})\right\},$$

$$p_2(\sigma) \propto \frac{1}{\sigma},$$

$g > 0$, and $(g, \bar{\boldsymbol{\beta}})$ denote hyperparameters. Information about the experiment is incorporated in $p_1(\boldsymbol{\beta}|\sigma)$ through $(\mathbf{X}'\mathbf{X})$. Inferences about $\boldsymbol{\beta}$ or σ in the above regression may be made from their marginal posterior densities based upon this prior.

The hyperparameter g may be assessed in many ways. One convenient way is to note that if

$$\boldsymbol{\beta} = (\beta_j), \qquad j = 1, \ldots, k,$$

$$\alpha \equiv \sum_{j=1}^{k} \operatorname{var}(\beta_j|\sigma) = g^{-1} \sum_{j=1}^{k} a_{jj},$$

where

$$A \equiv (X'X)^{-1} = (a_{ij}).$$

Thus, if we assess $\operatorname{var}(\beta_j|\sigma)$ for each j, and if we sum the assessments to get α, g is assessed as $\alpha^{-1}\Sigma_1^k a_{jj}$, where A is a known matrix fixed by the design of the regression (see Zellner, 1980 for discussion of an alternative assessment procedure based upon a "conceptual sample").

2.7.4 Assessing Complete Prior Distributions

Suppose I want to assess your complete prior distribution for an unobservable parameter θ, a scalar. In this case I want to assess more than just a couple of hyperparameters of a preassigned family of distributions which is mathematically convenient. (Assessment of prior distributions of vectors is treated in Section 5.3.1 for vague prior densities, and it is treated in Chapter IV for the more general case.) Below we suggest a four-step procedure that is useful in simple situations for assessing entire prior distributions.

ASSESSMENT STEPS

1. I ask you to give me a value $\theta_{.5}$ such that, in your view, it is equally likely that $\theta > \theta_{.5}$ or that $\theta < \theta_{.5}$. Furthermore, $\theta_{.5}$ is the median of your prior distribution.
2. Now suppose I tell you that $\theta > \theta_{.5}$. Then, I ask you to specify a value $\theta_{.75}$ such that it is equally likely that $\theta > \theta_{.75}$ or that $\theta < \theta_{.75}$. It is easy to check $\theta_{.75}$ is the 75th percentile of your prior distribution.
3. We can repeat step 2 for other fractiles greater than 50%. Then suppose I tell you that $\theta < \theta_{.5}$, and I ask you to specify a value $\theta_{.25}$ etc. In this manner we can assess any number of fractiles.
4. Combine the assessed fractiles by connecting them to form a smooth cdf curve. (The cdf is the cumulative distribution function.)

2.8 LIKELIHOOD PRINCIPLE

As presented in Section 2.2.3, the likelihood function of a set of observations $\mathbf{x}_1, \ldots, \mathbf{x}_N$ is their joint density (or probability mass function), when viewed as a function of the unknown parameter $\boldsymbol{\theta}$ which indexed the distribution from which the $\mathbf{x}_i$'s were generated. We denote it by $L(\mathbf{x}_1, \ldots, \mathbf{x}_N | \theta)$. Let $\mathbf{x} \equiv (\mathbf{x}_1, \ldots, \mathbf{x}_N)'$ and $\mathbf{y} \equiv (\mathbf{y}_1, \ldots, \mathbf{y}_M)'$ denote two distinct sets of observations, and suppose $L(\mathbf{x}|\boldsymbol{\theta}) \propto L(\mathbf{y}|\boldsymbol{\theta})$; that is, the likelihood functions for the two sets of observations are proportional (and the proportionality constant does not depend upon $\boldsymbol{\theta}$). Thus, their likelihood functions are the same, up to a multiplicative constant. But this means that maximum likelihood estimators of $\boldsymbol{\theta}$ will be the same, and if we adopt the same prior for $\boldsymbol{\theta}$, posterior inferences about $\boldsymbol{\theta}$ based upon $\mathbf{x}$ will be the same as those based upon $\mathbf{y}$. This fact is called the *likelihood principle*.

We note, along with Birnbaum (1962) that the evidence about θ arising from a given experiment and its observed outcome must be the same for two experiments whose associated likelihoods are proportional. We thus conclude, as a corollary to the likelihood principle, that all evidence about θ from an experiment and its observed outcome should be present in the likelihood function. If evidence about θ lies outside the likelihood function it is a violation of the likelihood principle.

For example, suppose that in n independent Bernoulli trials, r successes and $(n - r)$ failures are obtained. Let p denote the probability of success on a single trial. If the sampling takes place with n fixed, the resulting distribution for r, given p, is binomial. If, however, the sampling takes place with r held fixed, the resulting distribution for n, given p, is negative binomial. In either case, however, the likelihood function for p is proportional to $[p^r(1 - p)^{n-r}]$. Thus, the likelihood principle requires that all inferences about p be based upon the quantity in brackets, regardless of how the sampling was carried out. This idea that posterior inferences must be the same and that they must be based upon the likelihood function also follows from the ideas of coherence and is implied by Bayes' theorem. Thus, the likelihood principle requires that statistical inference be based upon only data found in the likelihood function. Principles of unbiasedness and admissibility which depend upon observations not yet taken (averages over the entire sample space) violate the likelihood principle. [For extended discussion of the implications and foundations of the likelihood principle, see Berger and Wolpert (1985).]

Likelihood Principle and Conditioning

Classical statistical inference is based upon pre-experimental evidence, that is, procedures are based upon their long run behavior over repeated experiments. The likelihood principle asserts that only post-experimental evidence should be used for inference, that is, inference should be based only upon observations actually collected in this experiment (and not upon observations that might have been taken, but weren't). Thus, we must condition upon the data from the current experiment.

As an example (see Berger and Wolpert, 1985, for an extensive discussion), suppose a substance to be analyzed can be sent either to a laboratory in New York, or a laboratory in California. The two labs seem to be equally competent so a fair coin is flipped to choose between them. The California lab is selected on the basis of the flip. Results of the analysis come back, and a conclusion must be reached. Should the conclusion take into account that the coin could have fallen the other way and the New York lab could have been chosen? The likelihood principle says no. Only the experiment actually performed should matter.

2.9 PREDICTIVE DISTRIBUTIONS; DE FINETTI'S THEOREM; PHILOSOPHY OF PREDICTIVISM

2.9.1 Predictive Distributions

Predictive distributions are concerned with predicting observables (no parameters are involved). [See, for example, Geisser (1980) for an extensive discussion of this philosophy of statistical inference; it is based upon de Finetti's theorem (discussed in Section 2.9.3).]

Definition

$\mathbf{X}_1, \ldots, \mathbf{X}_n$ denotes a sample from a distribution indexed by $\boldsymbol{\theta}$, with density $f(\mathbf{x} \mid \boldsymbol{\theta})$. For prior density $g(\boldsymbol{\theta})$, the posterior is given by

$$h(\boldsymbol{\theta} \mid \mathbf{x}_1, \ldots, \mathbf{x}_n) \propto \prod_1^n f(\mathbf{x}_j \mid \boldsymbol{\theta}) g(\boldsymbol{\theta}).$$

Suppose we wish to predict a new observation $\mathbf{y}$. *The predictive density for* $\mathbf{y}$ *is given by*

$$p(\mathbf{y} \mid \mathbf{x}_1, \ldots, \mathbf{x}_n) = \int f(\mathbf{y} \mid \boldsymbol{\theta}) h(\boldsymbol{\theta} \mid \mathbf{x}_1, \ldots, \mathbf{x}_n)\, d\boldsymbol{\theta}.$$

Intrinsic to these ideas is the fact that parameters can never be observed. [No one will ever be able to observe the mean of a $N(\theta, 1)$.] We therefore can never be certain how well such a construct as a mean has been estimated or approximated. What are much more relevant, however, are observables, that is, quantities capable of being observed. Philosophically, this means that in the field of statistics we should focus attention on observables and should therefore use predictive distributions as often as possible.

Predictive Density for Defective Parts Example

In the defective parts example, (see section 2.2.1) the likelihood for the old sample, y, is

$$L(y|\theta) = \theta^y(1-\theta)^{n-y},$$

and the posterior density with respect to a natural conjugate prior is

$$h(\theta|y) = \frac{\theta^{y+\alpha-1}(1-\theta)^{n-y+\beta-1}}{B(y+\alpha, n-y+\beta)}.$$

The likelihood of a new sample, z, is

$$L(z|\theta) = \binom{N}{z}\theta^z(1-\theta)^{N-z}.$$

So the predictive density of z, given y, is

$$p(z|y) = \int_0^1 \binom{N}{z}\theta^z(1-\theta)^{N-z}\frac{\theta^{y+\alpha-1}(1-\theta)^{n-y+\beta-1}}{B(y+\alpha, n-y+\beta)}\,d\theta$$

$$= \frac{\binom{N}{z}}{B(y+\alpha, n-y+\beta)}\int_0^1 \theta^{z+y+\alpha-1}(1-\theta)^{N-z+n-y+\beta-1}\,d\theta$$

$$= \frac{\binom{N}{z}B(z+y+\alpha, N-z+n-y+\beta)}{B(y+\alpha, n-y+\beta)}.$$

This is called the *probability mass function of a "beta binomial" distribution.*

2.9.2 Exchangeability

Definition

An infinite or potentially infinite sequence of random variables $\theta_1, \ldots, \theta_q, \ldots$ is said to be exchangeable if the joint distribution of any finite subset of the

$\boldsymbol{\theta}_i$'s is invariant under permutations of the subscripts. The concept has been extended to strictly finite sequences, as well. A special case occurs when $\boldsymbol{\theta}_i$'s are i.i.d. Then they are certainly exchangeable (but exchangeable random variables are not necessarily independent).

Example of (Multivariate) Exchangeability Without Independence
Let $\boldsymbol{\theta}_j:(p \times 1)$, and

$$\underset{(pq \times 1)}{\boldsymbol{\theta}} \equiv \begin{pmatrix} \boldsymbol{\theta}_1 \\ \vdots \\ \boldsymbol{\theta}_q \end{pmatrix}, \qquad \boldsymbol{\theta} \sim N(\boldsymbol{\phi}, \boldsymbol{\Phi}),$$

where $\mathbf{a}:(p \times 1)$, and

$$\boldsymbol{\phi} = \mathbf{e}_q \otimes \mathbf{a} = \begin{pmatrix} \mathbf{a} \\ \vdots \\ \mathbf{a} \end{pmatrix}, \qquad \boldsymbol{\Phi} = \begin{pmatrix} \boldsymbol{\Omega} & & \boldsymbol{\Psi} \\ & \ddots & \\ \boldsymbol{\Psi} & & \boldsymbol{\Omega} \end{pmatrix}, \boldsymbol{\Omega}:(p \times p), \boldsymbol{\Psi}:(p \times p),$$

$\boldsymbol{\Phi} > 0$, $\boldsymbol{\Omega} > 0$, $\boldsymbol{\Psi} > 0$, $\mathbf{e}_q \equiv (1, \ldots, 1)'$. The operation $\otimes$ means direct product of two matrices or vectors. The operation is defined mathematically at the end of Section 6.5. Note that $\boldsymbol{\Phi} > 0$ means that $\boldsymbol{\Phi}$ is a positive definite, symmetric matrix. Thus, $\boldsymbol{\Phi}$ is a matrix intraclass covariance matrix, and the $\boldsymbol{\theta}_i$'s are exchangeable (finitely) but are not independent. In an intraclass covariance matrix, the diagonal elements (the variances) are the same, and the off-diagonal elements (the covariances) are the same.

Some relevant discussions of exchangeability are given in de Finetti, 1937, [illegible], Hewitt and Savage, 1955; Savage, 1954, Section 3.7; Lindley and Novick, 1982; Diaconis, 1977, 1980; Koch and Spizzichino, 1982; and Diaconis and Freedman, 1980. Bayesian inference using exchangeability in the multivariate normal distribution is discussed in Dickey, Lindley, and Press (1985).

2.9.3 de Finetti's Theorem

In 1937, de Finetti showed that infinite sequences of exchangeable binary events having any joint distribution can be thought of as mixtures of coin-tossing experiments, that is, mixtures of i.i.d. binary random variables (i.e., Bernoulli trials) (de Finetti, 1937). The theorem asserts that if x_j is a 0-1 random variable that denotes the outcome of coin toss j, the predictive probability of a given sequence $\{x_1, \ldots, x_n\}$ that yields a total of r

successes in n trials is

$$p\{x_1,\ldots,x_n|r,n\} = \int_0^1 \prod_{i=1}^{n} \theta^{x_i}(1-\theta)^{1-x_i}\, dF(\theta)$$

$$= \int_0^1 \theta^r(1-\theta)^{n-r}\, dF(\theta),$$

where $F(\theta)$ denotes some proper cdf on (0, 1). Thus, the right-hand side is a mixture of binomial probabilities, and $F(\theta)$ is the mixing distribution. That is, the predictive probability distribution of the exchangeable random variables $X_1,\ldots, X_n$ may be obtained by acting as if the X's were independent, conditional on θ, and identically distributed and as if their density were averaged over θ. The $dF(\theta)$ in the integral should be interpreted as $f(\theta)\, d\theta$, where $f(\theta)$ is the density corresponding to $F(\theta)$, in the case where θ is continuous; and if θ is discrete the integral should be interpreted as a sum, and $f(\theta)$ is then a probability mass function. The integral, as written, which combines cases of both continuous and discrete θ, is called a *Stieltjes integral*.

An equivalent, intuitively appealing statement of the de Finetti theorem is the following:

Theorem. Let S_n denote the number of successes in n exchangeable Bernoulli trials in which you don't know the probability of success on a single trial, θ. Then:

1. $P\{S_n = r\} = \int_0^1 \binom{n}{r}\theta^r(1-\theta)^{n-r}\, dF(\theta)$, and
2. $\lim_{n\to\infty} S_n/n = \theta$ with probability one, for any mixing distribution function of θ, $F(\theta)$.

The choice made by Bayes and Laplace for the mixing distribution was the uniform distribution. That is, all values of θ are equally likely (that's what we get from Laplace's principle of insufficient reason). In this case, $dF(\theta) = d\theta$, and for $r = 0, 1,\ldots, n$,

$$P\{S_n = r\} = \int_0^1 \binom{n}{r}\theta^r(1-\theta)^{n-r}\, d\theta = \frac{1}{n+1}$$

That is, for 9 trials, for example, the probability of zero successes is the same as that for one success, or two successes, or . . . , or 9 successes—namely, 10%.

Two special cases of de Finetti's theorem of particular interest are:

1. The distribution of θ is discrete, and concentrated at $\theta = \theta_0$ $(P\{\theta = \theta_0\} = 1)$. Then, de Finetti's theorem asserts that:

$$P\{S_n = r\} = \binom{n}{r}\theta_0^r(1-\theta_0)^{n-r}.$$

   That is, if $\theta = \theta_0$ with probability one, then we get the usual binomial formula for the probability of r successes in n trials, conditional on $\theta = \theta_0$.
2. When the distribution of θ is discrete and concentrated at $\theta = \theta_0$, the second part (the limit part) of de Finetti's theorem gives, for binary outcomes $X_1, \ldots, X_n$,

$$\lim_{n\to\infty} \frac{1}{n}\sum_{i=1}^{n} X_i = \theta, \text{ with probability one.}$$

   But $P\{S_1 = 1\} = P\{X_1 = 1\} = EX_1 = \binom{1}{1}\theta^1(1-\theta)^0 = \theta.$

So the strong law of large numbers results as a special case of de Finetti's theorem, from exchangeability.

The de Finetti result may be expressed in more general form by considering general (not necessarily binary) random variables $X_1, \ldots, X_N, \ldots$ which are exchangeable; otherwise, we only need to specify that they are jointly distributed and that any finite number of these have the joint density $h(x_1, \ldots, x_N)$. The de Finetti theorem then asserts that $h(\cdot)$ may be represented as the mixture

$$h(\mathbf{x}_1, \ldots, \mathbf{x}_N) = \int f(\mathbf{x}_1|\boldsymbol{\theta}) \cdots f(\mathbf{x}_N|\boldsymbol{\theta})\, dF(\boldsymbol{\theta}),$$

and $F(\cdot)$ is again the cdf of the mixing distribution. Thus, the predictive density for the exchangeable random variables can be thought of as a mixture of i.i.d. random variables with density $f(\cdot \mid \cdot)$.

Note that de Finetti's theorem requires that there be an infinite (or potentially infinite) number of exchangeable events (or random variables). It has been shown, however, that while de Finetti's theorem does not hold exactly for finitely exchangeable events, the result does hold approximately for sufficiently large but finite sequences (for details, see Diaconis, 1977; also see Diaconis and Freedman, 1980).

Another implication of the theorem is that when we have i.i.d. random variables for the sampling density and also have a known prior distribution, the unconditional distribution is exchangeable.

We note that de Finetti's theorem makes it possible to use the subjective (personalistic) interpretation of statements involving an unknown probability. For example, in a conventional dice game in which a "seven" seems to come up very frequently, you might specify that you believe that the probability of the outcome "seven," θ, is uniformly distributed over the range $1/4 \leq \theta \leq 3/4$. Then, de Finetti's theorem could be used to evaluate the chances of, say five straight sevens, that is,

$$\int_{1/4}^{3/4} 2\theta^5 \, d\theta.$$

Implicit in your belief is, of course, the notion that trials are exchangeable.

Some additional interesting interpretations of de Finetti's theorem are given in Skyrms (1984).

2.10 ROBUST AND NONPARAMETRIC BAYESIAN INFERENCE: THE BAYESIAN BOOTSTRAP

It is frequently the case that statistical inference from a set of data is very sensitive to the assumptions made about the data and the model representing the data. When there are large quantities of data, such sensitivity tends to be minimal (Box and Tiao, 1973 refer to this property as *criterion robustness*); however, with moderate or small data bases, inferences tend to be strongly model dependent. If we change the likelihood somewhat, how much do our inferences change? If we change the prior somewhat, how much do our posterior inferences change? (If our posterior inferences don't change very much Box and Tiao, 1973 call it *inference robustness*.) This type of concern has led to a striving for statistical procedures which require that we make only very few assumptions. Such procedures are called *robust* (see e.g., Huber, 1977; and Kadane, 1984).

Jeffreys (1961) suggested that we can hedge against uncertainties about the tail behavior of our posterior distribution by adopting families of prior distributions that have (flat-tail) Cauchy distribution behavior. (This will imply using estimators that are nonlinear functions of the MLE.) Such a prior will make the posterior more sensitive to outliers.

In a sampling theory context, Efron (1979, 1982) proposed the "Bootstrap." This method, in its original form, does not assume any parametric form for the data sampling distribution. Instead, Efron suggests that we should: (1) form the empirical cdf from the sample of size N; (2) take many samples (with replacement) from this empirical distribution (these samples are called the *bootstrap samples*); and (3) use these samples to find the distribution of an estimator.

For example, if we want to estimate the variance of an estimator of θ, say $\hat{\theta}$, we take a bootstrap sample from the empirical cdf* and form $\hat{\theta}_1$; then we take another bootstrap sample and form $\hat{\theta}_2$; and so on. Thus, we form K values of $\hat{\theta}$ from K bootstrap samples of size N. Now form the bootstrap estimate of the variance of $\hat{\theta}$,

$$\hat{\sigma}^2 = \frac{1}{K-1} \sum_{j=1}^{K} \left(\hat{\theta}_j - \bar{\theta}\right)^2,$$

where $\bar{\theta} = (1/K)\Sigma_{j=1}^{K}\hat{\theta}_j$.

Note that if $\hat{\theta}$ is a sample mean, finding an estimator of the variance in the usual way is straightforward (as the sample variance divided by the sample size). If, however, $\hat{\theta}$ is a sample median, or something more complicated, an estimator of variance is harder to find.

The bootstrap procedure is intuitively sensible, and it converges to the true value in large samples. In this respect it is analogous to maximum likelihood estimation. In both cases, however, we are uncertain about the small sample behavior of the procedure.

The statistician who wishes to adopt a robust Bayesian posture can proceed with a *Bayesian bootstrap*. [A critical position is taken by Rubin (1981).] Such an approach is the following. Suppose we have a sample $X_1, \ldots, X_N$ from some distribution with unknown parameter θ. For concreteness we take θ to be the mean, but θ could be any parameter. Now form the empirical cdf, $F_N(x)$, and draw K bootstrap samples of size N independently from $F_N(x)$, with replacement. Form $\hat{\theta} = \bar{X}$ for each bootstrap sample, yielding $\hat{\theta}_1, \ldots, \hat{\theta}_K$. Now form the empirical density of the K $\hat{\theta}$'s, and call it $f_N(\hat{\theta}|\theta)$. One method of forming such an empirical density is presented in Section 4.5. [See also Lo (1984) and Silverman (1986).] Note that density estimation typically requires that the $\hat{\theta}_1, \ldots, \hat{\theta}_K$ be i.i.d. Since these variables are not independent, the procedure is only an approximation. Let $p(\theta)$ denote the prior density of θ. The empirical posterior density we use is

$$h_N(\theta|X_1, \ldots, X_N) \propto f_N(\hat{\theta}|\theta)p(\theta).$$

*The empirical cdf $F_N(x)$ for i.i.d. observations $X_1, \ldots, X_N$ from some unknown cdf $F(x)$ is given by $F_N(x) \equiv$ (Number of X_i's $\leq x)/N$.

The normalizing constant may be obtained by numerical integration. For large N, this density will approach the result obtainable from Bayes' theorem if we were to take a parametric approach, and if we knew the likelihood function. Thus, inferences made from $h_N(\theta | X_1, \ldots, X_N)$ can reasonably be thought of as approximate Bayesian inferences. Monte Carlo simulations of the behavior of these types of estimators have been carried out by Boos and Monahan (1983). They have found that in moderate or large samples, the bootstrap estimators behave as Bayes' estimators.

2.11 EXAMPLE: PATERNITY SUITS

Over the years, suits have appeared in courts, where paternity is alleged. The defense has successfully used Bayesian inference to negate paternity based upon blood tests and unlikely phenotypic blood types. New biological methods of discriminating among alleged fathers based upon genes and DNA matching, and not blood types, promise to revolutionize this whole field with enormously increased accuracy, although the statistical methodology will remain the same as described here (see, for example, Lewis, 1988).

For simplicity we consider a (phenotypic) partition of the population into just four main blood types: O, AB, A, and B, with population percentages:

Type	O	AB	A	B
%	45	4	41	10

More complex partition by blood types would include more than four types.

Observational Data

Our observational data consists of a triple of observations representing the blood phenotypes of the mother, child, and alleged father (x, y, z), where x, y, and z are categorical variables, each of which has four possible values ($x = 1$ if the mother has phenotype O; $x = 2$ if she has phenotype AB; etc.); similarly for y and z for the child and alleged father. Define the likelihood ratio for the genetic evidence:

$$\tilde{L} = \frac{P\{(x, y, z) | \text{Paternity}\}}{P\{(x, y, z) | \text{Nonpaternity}\}}.$$

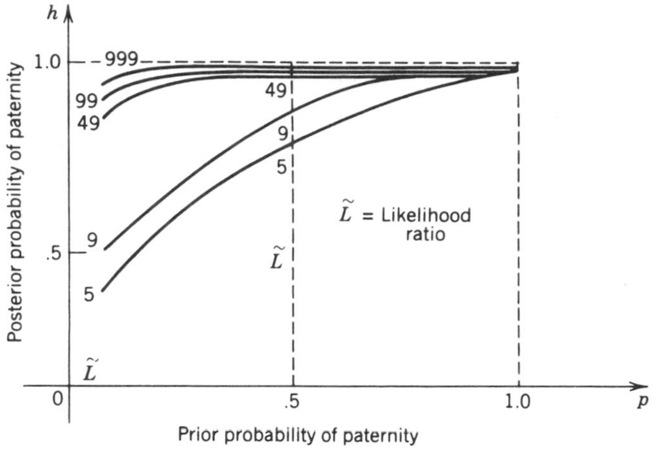

Figure 2.6. Implications of Bayes' theorem in resolving paternity suits.

The likelihoods must be determined by means of biological tests, which are not always straightforward. Define p = prior probability of paternity (nongenetic). Bayes' theorem gives

$$h = P\{\text{Paternity}|(x, y, z)\} = \frac{p\tilde{L}}{p\tilde{L} + (1 - p)}.$$

Under the assumption of nonpaternity, we assume independence of the child's phenotype with that of the father's, so that

$$\begin{aligned} P\{(x, y, z)|\text{Nonpaternity}\} \\ &= [P\{(x, y)|\text{Nonpaternity}\}][P\{z|\text{Nonpaternity}\}] \\ &= [P\{(x, y)\}][P\{z\}]. \end{aligned}$$

For this genetic system (or any other that might be used) we sum over all mother–child phenotype combinations to compute $\tilde{L}$, based upon genetic laws (which are not obeyed perfectly). If we use several genetic systems, we multiply the corresponding $\tilde{L}$'s to obtain an overall $\tilde{L}$.

The implications of Bayes' theorem for various combinations of $\tilde{L}$ and p are shown in Figure 2.6.

In the courts, the judges will typically accept a prior probability of $p = .5$, in the interest of not prejudging an individual. Then,

$$h = \frac{\tilde{L}}{1 + \tilde{L}}.$$

It is found in some cases that this type of analysis will yield an extremely high or extremely low probability of paternity, which is compelling to a judge or jury. Whether a $p = .5$ is reasonable is another important issue. Some argue that p should be higher for an accused father. In any case, a posterior that is fairly insensitive to p is a strong argument.

For additional discussion of the use of Bayesian inference in the study of paternity see, for example, Berry and Geisser (1986).

2.12 SUMMARY: ADVANTAGES OF BAYESIAN APPROACH

The advantages of using the Bayesian approach are as follows:

1. We can take our earlier practical experience into account explicitly.
2. We often obtain shorter confidence (credibility) intervals using proper prior distributions than we could obtain if we ignored our practical experience.
3. We often obtain estimators (by using proper priors to incorporate prior experience) with smaller variance or mean-squared error than we could have obtained had we ignored our practical experience.
4. We often obtain more accurate forecasts and predictions with proper prior distributions.
5. Proper prior distribution unique Bayesian procedures are admissible.
6. We can apply Bayesian methods in situations where objective methods are needed, by using vague priors. Often, frequentist analogs will result. This is one approach we recommend for public policy situations; another is to provide a mapping of the various posterior probabilities that result from various priors.
7. We can test hypotheses without predetermining the outcome of the test according to the selection of the sample size, and we need not prespecify an arbitrary level of significance.

EXERCISES

2.1 Explain the importance of Bayesian estimators with respect to proper prior distributions in decision theory.

2.2 Explain the use of a "highest posterior probability" credibility interval.

2.3 Compare credibility and confidence intervals.

2.4 What is meant by "Lindley's paradox"? (See Section 2.3.2.)

2.5* What is meant by lower and upper probability (see Dempster, 1967)?

2.6* What is meant by belief functions (see Shafer, 1976), and how do they generalize Bayesian inference?

2.7 Explain the meaning of the statement "Most scientific inference is already partly subjective, and it always has been."

2.8 Explain Laplace's "principle of insufficient reason."

2.9 Let θ denote the expected number of live chickens in California on a given day. Explain how you would formulate your prior distribution for θ.

2.10 Explain the "closure property" for natural conjugate prior distributions.

2.11 What is meant by the "likelihood principle"?

2.12 Suppose $X|\theta \sim N(\theta, 1)$, and the prior density for θ is $N(0, 1)$. Give the predictive density for a new observation (see Section 2.9.1).

2.13 What is meant by "empirical Bayes estimation"? (See Section 2.7.1.)

2.14 What is meant by a "public policy prior," which is also called an "objective prior"?

2.15 If X, given λ, follows a Poisson distribution, give a natural conjugate prior density for λ.

2.16 What is meant by a "Bayesian bootstrap," and under what conditions might you be interested in using one?

2.17 Suppose $\mathbf{X}: (p \times 1)$, and $\mathbf{X} \sim N(\boldsymbol{\theta}, \mathbf{I})$. Find a natural conjugate prior family for $\boldsymbol{\theta}$.

2.18 Suppose $X \sim N(\theta, 1)$ and we take three independent observations, $X_1 = 2$, $X_2 = 3$, $X_3 = 4$. If we adopt a vague prior for θ, then:

(a) Find the posterior distribution for θ given the data.

(b) Find a two-tailed 95% credibility interval for θ.

2.19 Explain why "unbiasedness" is a property that violates the spirit of the likelihood principle.

2.20 Why is "admissibility" not usually an important criterion for a Bayesian statistician for choosing estimators which have good decision-theoretic properties?

2.21 Suppose r denotes the number of successes in n trials, and r follows a binomial distribution with parameter p. Carry out a Bayesian test of the hypothesis H: $p = .2$ vs. the alternative A: $p = .8$, where these are the only two possibilities. Assume that $r = 3$ and $n = 10$ and that the prior probabilities of H and A are equal.

*Asterisked exercises require reference to sources outside of this text. Full reference information can be found in the bibliography at the back of this book.

CHAPTER III

Approximations, Numerical Methods, and Computer Programs

3.1 INTRODUCTION

The Bayesian paradigm is conceptually simple, intuitively plausible, and probabilistically elegant. Its numerical implementation is not always easy and straightforward, however. Posterior distributions are often expressible only in terms of complicated analytical functions; we often know only the kernel of the posterior density (and not the normalizing constant); we often can't readily compute the marginal densities and moments of the posterior distribution in terms of exact closed form explicit expressions; and it is sometimes difficult to find numerical fractiles of the posterior cdf. We quote Dempster (1980, p. 273):

> The application of inference techniques is held back by conceptual factors and computational factors. I believe that Bayesian inference is conceptually much more straightforward than non-Bayesian inference, one reason being that Bayesian inference has a unified methodology for coping with nuisance parameters, whereas non-Bayesian inference has only a multiplicity of ad hoc rules. Hence, I believe that the major barrier to much more widespread application of Bayesian methods is computational.... The development of the field depends heavily on the preparation of effective computer programs.

In this chapter we present a summary of the leading methods that have been developed to attack the computational problems of implementing Bayesian inference. We begin with the large sample (normal) approximation to the posterior distribution (Section 3.2); we move on to the various analytical approximation methods that have been proposed for evaluating not only the integrals that arise naturally in Bayesian inference, but also the integrals for the normalizing constant, for the marginal posterior densities, for the posterior moments, and for fractiles of the posterior distribution (Section 3.3). In Section 3.4 we show how to use simulation methods to study the behavior of the sometimes quite complicated posterior distributions that arise in Bayesian inference. Finally, in Section 3.5 we provide a summary of some available Bayesian computer software programs.

3.2 LARGE-SAMPLE POSTERIOR DISTRIBUTIONS

One method of dealing with the practical problems of implementing the Bayesian paradigm, which can be useful when moderate or large samples are available, is to adopt the limiting large-sample normal distribution for posterior inferences. (It is often a surprisingly good approximation even in small samples.) This limiting distribution is straightforward to determine, and the regularity conditions required for its existence are usually satisfied. The result is given in the theorem below.

Theorem 3.2.1. Let $\mathbf{x}_1, \ldots, \mathbf{x}_n$ denote p-variate observations from a sampling distribution with joint density $f(\mathbf{x}_1, \ldots, \mathbf{x}_n|\boldsymbol{\theta})$, $\boldsymbol{\theta}: (k \times 1)$, and denote the prior density for $\boldsymbol{\theta}$ by $g(\boldsymbol{\theta}) > 0$. Under suitable regularity conditions, the limiting posterior distribution of $\boldsymbol{\theta}$ conditional on the data is given by

$$\lim_{n \to \infty} \mathscr{L}\left\{\hat{\boldsymbol{\Sigma}}^{-1/2}(\boldsymbol{\theta} - \hat{\boldsymbol{\theta}})|\mathbf{x}_1, \ldots, \mathbf{x}_n\right\} = N(\mathbf{0}, \mathbf{I}_k),$$

where $\hat{\boldsymbol{\theta}}$ denotes the maximum likelihood estimator of $\boldsymbol{\theta}$, $\hat{\boldsymbol{\Sigma}}^{-1} \equiv \hat{\boldsymbol{\Lambda}} \equiv (\hat{\lambda}_{ij})$, and

$$\hat{\lambda}_{ij} = -\left.\frac{\partial^2 \log f(\mathbf{x}_1, \ldots, \mathbf{x}_n|\boldsymbol{\theta})}{\partial \theta_i \, \partial \theta_j}\right|_{\boldsymbol{\theta} = \hat{\boldsymbol{\theta}}},$$

$\mathbf{I}_k$ denotes the identity matrix of order k, and N denotes the normal distribution.

REMARK 1: An equivalent statement of the result in the theorem is that if

$$\boldsymbol{\phi} \equiv \hat{\boldsymbol{\Sigma}}^{-1/2}(\boldsymbol{\theta} - \hat{\boldsymbol{\theta}}) \equiv (\phi_i),$$

then

$$\lim_{n\to\infty} P\{\phi_1 \le t_1, \ldots, \phi_k \le t_k | \mathbf{x}_1, \ldots, \mathbf{x}_n\}$$
$$= \int_{-\infty}^{t_1} \cdots \int_{-\infty}^{t_k} \frac{1}{(2\pi)^{k/2}} e^{-\boldsymbol{\phi}'\boldsymbol{\phi}/2}\, d\boldsymbol{\phi}.$$

REMARK 2: Note that for the theorem to hold, $g(\boldsymbol{\theta})$ must be strictly positive (and continuous) at all interior points of the parameter space (otherwise, no amount of sample data could revise the prior belief that $g(\boldsymbol{\theta}_0) = 0$ for some $\boldsymbol{\theta}_0$). Moreover, the likelihood function $f(\mathbf{x}_1, \ldots, \mathbf{x}_n | \boldsymbol{\theta})$ must be twice differentiable at $\hat{\boldsymbol{\theta}}$, so that $\hat{\boldsymbol{\Sigma}}$ will exist; of course we must also have $\hat{\boldsymbol{\Sigma}}$ nonsingular, so that $\hat{\boldsymbol{\Lambda}}$, the precision matrix, will also exist. We assume also that $\boldsymbol{\theta}$ is not one of the bounds of the support of $\mathbf{x}_i$.

REMARK 3: Note that the large sample posterior distribution of $\boldsymbol{\theta}$ does not depend upon the prior; that is, in large samples, the data totally dominate the prior beliefs.

Proof: See Le Cam (1956, p. 308). It may also be shown that when the sample data is i.i.d., the required regularity conditions are the same as those required for proving asymptotic normality of the MLE (see Hyde and Johnstone, 1979). When the observations are correlated (such as in time-series data, or in spatially correlated data arising in geophysics), the required regularity conditions are more complicated. We note that the result in this theorem arises from a Taylor-series expansion of the log-likelihood function about $\hat{\boldsymbol{\theta}}$. For a convenient, heuristic proof, see Lindley, 1965, Vol. 2. □

Example 3.2.1. Let $x_1, \ldots, x_n$ denote the numbers of small aircraft arriving at the municipal airport in Riverside, California during time intervals $t_1, \ldots, t_n$. Assume the x_i's are mutually independent, and suppose the probability mass function of x_j is Poisson and is given by

$$l(x_j|\theta) = \frac{e^{-\theta}\theta^{x_j}}{x_j!}, \qquad \theta > 0,$$

so that

$$f(x_1, \ldots, x_n|\theta) = \prod_1^n \frac{e^{-\theta}\theta^{x_j}}{x_j!} = \frac{e^{-n\theta}\theta^{\Sigma_1^n x_j}}{\Pi_1^n x_j!}.$$

The maximum likelihood estimator of θ is found as follows. The log-likelihood of θ is

$$L(\theta) \equiv \log f(x_1, \ldots, x_n | \theta)$$

$$= -n\theta + (\log \theta) \sum_1^n x_j - \log\left(\prod_1^n x_j! \right).$$

$$\frac{dL(\theta)}{d\theta} = -n + \frac{\sum_1^n x_j}{\theta}.$$

$$\frac{dL(\theta)}{d\theta} = 0 \text{ implies } \hat{\theta} = \bar{x} = \frac{1}{n} \sum_1^n x_j.$$

The second derivative of $L(\theta)$ with respect to θ, evaluated at $\hat{\theta}$, is

$$-\hat{\Lambda} \equiv \left. \frac{d^2 L(\theta)}{d\theta^2} \right|_{\theta = \hat{\theta}} = -\left. \frac{1}{\theta^2} \sum_1^n x_j \right|_{\theta = \hat{\theta}},$$

or

$$\hat{\Sigma}^{-1} = \hat{\Lambda} = \frac{1}{\hat{\theta}^2} \sum_1^n x_j = \frac{n}{\bar{x}}.$$

Suppose the prior distribution of θ is log-normal with density

$$g(\theta) = \frac{1}{\theta \sigma_0 \sqrt{2\pi}} e^{-(1/2\sigma_0^2)(\log \theta - \mu)^2}, \qquad \theta > 0,$$

where (μ, σ_0^2) are known parameters that are assessed. The posterior density then becomes

$$h(\theta | x_1, \ldots, x_n) \propto \left[\frac{1}{\theta} e^{-(1/2\sigma_0^2)(\log \theta - \mu)^2} \right] \left[e^{-n\theta} \theta^{\sum_1^n x_j} \right]$$

$$h(\theta | x_1, \ldots, x_n) \propto \theta^{\sum_1^n x_j - 1} e^{-n\theta - (2\sigma_0^2)^{-1}(\log \theta - \mu)^2}, \qquad \theta > 0.$$

This posterior distribution is quite complicated; we cannot readily evaluate the moments, and we do not even know the proportionality constant. If n is sufficiently large, however, we can rely on the large-sample result in Theorem 3.2.1. Since for this example, $\hat{\Sigma}^{-1} = n/\bar{x}$, we find that for large n, we have the approximate distributional result

$$\mathcal{L}\left\{ \sqrt{\frac{n}{\bar{x}}} (\theta - \bar{x}) | x_1, \ldots, x_n \right\} \simeq N(0,1).$$

Equivalently, for large n, we have

$$\mathscr{L}\{\theta|x_1,\ldots,x_n\} \simeq N\left(\bar{x}, \frac{\bar{x}}{n}\right),$$

a much simpler result from which to make inferences than the exact posterior distribution.

3.3 APPROXIMATE EVALUATION OF BAYESIAN INTEGRALS

To implement the Bayesian paradigm it is necessary to be able to evaluate ratios of integrals of the form

$$I(\mathbf{x}_1,\ldots,\mathbf{x}_n) \equiv \frac{\int u(\boldsymbol{\theta})e^{L(\boldsymbol{\theta})+\rho(\boldsymbol{\theta})}\,d\boldsymbol{\theta}}{\int e^{L(\boldsymbol{\theta})+\rho(\boldsymbol{\theta})}\,d\boldsymbol{\theta}}, \tag{3.3.1}$$

where $L(\boldsymbol{\theta}) \equiv \log f(\mathbf{x}_1,\ldots,\mathbf{x}_n|\boldsymbol{\theta}) = \log \prod_1^n l(\mathbf{x}_j|\boldsymbol{\theta})$ denotes the log of the likelihood function, $\rho(\boldsymbol{\theta}) \equiv \log g(\boldsymbol{\theta})$ denotes the log of the prior density, and $u(\boldsymbol{\theta})$ is an arbitrary function of $\boldsymbol{\theta}$.

For example, if θ is one-dimensional, $u(\theta) = \theta$ is the mean of the posterior distribution (the Bayesian estimator of θ for quadratic loss); $u(\theta) = \theta^k$ more generally yields the kth moment of the posterior distribution. (Note that $\boldsymbol{\theta}$ may be multidimensional.) Since $I(\mathbf{x}_1,\ldots,\mathbf{x}_n) \equiv E[u(\boldsymbol{\theta})|\mathbf{x}_1,\ldots,\mathbf{x}_n]$, $I(\cdot)$ provides the posterior mean of the arbitrary function $u(\boldsymbol{\theta})$. Evaluating the denominator integral in Eq. (3.3.1) yields the normalizing constant in the posterior density.

Marginal posterior densities may be obtained from a variant of the ratio of integrals in Eq. (3.3.1). Partition $\boldsymbol{\theta}$, letting $\boldsymbol{\theta} \equiv (\dot{\boldsymbol{\theta}}', \ddot{\boldsymbol{\theta}}')'$, let $h(\cdot)$ denote the posterior density, and define

$$\begin{aligned} I^*(\dot{\boldsymbol{\theta}};\mathbf{x}_1,\ldots,\mathbf{x}_n) &\equiv \int h(\dot{\boldsymbol{\theta}},\ddot{\boldsymbol{\theta}}|\mathbf{x}_1,\ldots,\mathbf{x}_n)\,d\ddot{\boldsymbol{\theta}} \\ &= \int h(\boldsymbol{\theta}|\mathbf{x}_1,\ldots,\mathbf{x}_n)\,d\ddot{\boldsymbol{\theta}} \\ &= \frac{\int e^{L(\boldsymbol{\theta})+\rho(\boldsymbol{\theta})}\,d\ddot{\boldsymbol{\theta}}}{\int e^{L(\boldsymbol{\theta})+\rho(\boldsymbol{\theta})}\,d\boldsymbol{\theta}}. \end{aligned} \tag{3.3.2}$$

Note that the ratio of integrals in Eq. (3.3.2) is the marginal posterior density of $\dot{\boldsymbol{\theta}}$, given the data.

In Section 3.3.1 below, we present three of the analytical approximations that have been proposed for evaluating $I(\cdot)$ and $I^*(\cdot)$ in Eqs. (3.3.1) and (3.3.2). They differ from one another in their scope of applicability and in their accuracy. They all improve on the large-sample normal approximation presented in Section 3.2 by not requiring the sample size to be so large.

3.3.1 Low Dimension (by Numerical Approximation Methods)

The Lindley Approximation

The first approximation, developed in Lindley (1980), is given in the following theorem. It is most useful when the dimension (number of parameters, p) is low (say, ≤ 5).

Theorem 3.3.1. For n sufficiently large, so that $L(\boldsymbol{\theta})$ defined in Eq. (3.3.1) concentrates around a unique maximum likelihood estimator $\hat{\boldsymbol{\theta}} \equiv \hat{\boldsymbol{\theta}}(\mathbf{x}_1, \ldots, \mathbf{x}_n)$, for $\boldsymbol{\theta} \equiv (\theta_i)$: $p \times 1$, $\hat{\boldsymbol{\theta}} = (\hat{\theta}_i)$, it follows that $I(\cdot)$, defined in Eq. (3.3.1), is expressible approximately as

$$I(\mathbf{x}_1, \ldots, \mathbf{x}_n) \simeq u(\hat{\boldsymbol{\theta}}) + \frac{1}{2} \sum_{i=1}^{p} \sum_{j=1}^{p} \left[\left. \frac{\partial^2 u(\boldsymbol{\theta})}{\partial \theta_i \, \partial \theta_j} \right|_{\boldsymbol{\theta}=\hat{\boldsymbol{\theta}}} + \left\{ 2 \left. \frac{\partial u(\boldsymbol{\theta})}{\partial \theta_i} \right|_{\boldsymbol{\theta}=\hat{\boldsymbol{\theta}}} \right\} \left\{ \left. \frac{\partial \rho(\boldsymbol{\theta})}{\partial \theta_j} \right|_{\boldsymbol{\theta}=\hat{\boldsymbol{\theta}}} \right\} \right] \hat{\sigma}_{ij}$$
$$+ \left\{ \frac{1}{2} \sum_{i=1}^{p} \sum_{j=1}^{p} \sum_{k=1}^{p} \sum_{l=1}^{p} \left. \frac{\partial^3 L(\boldsymbol{\theta})}{\partial \theta_i \, \partial \theta_j \, \partial \theta_l} \right|_{\boldsymbol{\theta}=\hat{\boldsymbol{\theta}}} \right\} \cdot \left\{ \left. \frac{\partial u(\boldsymbol{\theta})}{\partial \theta_k} \right|_{\boldsymbol{\theta}=\hat{\boldsymbol{\theta}}} \right\} \hat{\sigma}_{ij} \hat{\sigma}_{kl}, \tag{3.3.3}$$

where $\hat{\sigma}_{kj}$ denotes the (i, j) element of $\hat{\boldsymbol{\Sigma}} \equiv (\hat{\sigma}_{ij})$, for $\hat{\boldsymbol{\Sigma}}^{-1} = \hat{\boldsymbol{\Lambda}} = (\hat{\lambda}_{ij})$, and

$$\hat{\lambda}_{ij} = - \left. \frac{\partial^2 L(\boldsymbol{\theta})}{\partial \theta_i \, \partial \theta_j} \right|_{\boldsymbol{\theta}=\hat{\boldsymbol{\theta}}}.$$

Proof: See Lindley (1980).

REMARK 1: The first term in Eq. (3.3.3) is $O(1)$; the other terms are all $O(1/n)$ and are called *correction terms*. The overall approximation in the theorem is $O(1/n)$, so the first term neglected is $O(1/n^2)$.

REMARK 2: The approximation in eq. (3.3.3) involves the third derivative of $L(\boldsymbol{\theta})$. The last term, however, does not involve the prior at all, and only the first derivative of the prior is involved. Interestingly, by contrast with the large-sample approximation in Section 3.2, which depended strictly on the form of the likelihood function, this approximation depends upon the form of the prior as well.

Example 3.3.1. We reconsider Example 3.2.1, but now we focus attention on the Bayesian estimator for a quadratic loss function, the posterior mean. In that example, we saw that the large-sample (first-order) approximation to the posterior mean is the MLE, namely, $\hat{\theta} = \bar{x}$. Now we use Theorem 3.3.2 to improve upon the approximation and to see the effect of the prior on the posterior mean.

First simplify Eq. (3.3.3) for the case of a single one-dimensional parameter of interest and take $p = 1$. This gives

$$I(x_1, \ldots, x_n) \simeq u(\hat{\theta}) + \frac{1}{2}\left\{ \left.\frac{\partial^2 u(\theta)}{\partial\theta^2}\right|_{\theta=\hat{\theta}} + 2\left.\frac{\partial u(\theta)}{\partial\theta}\frac{\partial\rho(\theta)}{\partial\theta}\right|_{\theta=\hat{\theta}} \right\}\hat{\sigma}^2$$

$$+ \left\{ \left.\frac{1}{2}\frac{\partial^3 L(\theta)}{\partial\theta^3}\frac{\partial u(\theta)}{\partial\theta}\right|_{\theta=\hat{\theta}} \right\}\hat{\sigma}^4. \tag{3.3.4}$$

Next take $u(\theta) = \theta$ (so that $\partial u(\theta)/\partial\theta = 1$ and $\partial^2 u(\theta)/\partial\theta^2 = 0$), and recall that $\hat{\theta} = \bar{x}$. Substituting into Eq. (3.3.4) gives

$$I(x_1, \ldots, x_n) \simeq \bar{x} + \left\{ \left.\frac{\partial\rho(\theta)}{\partial\theta}\right|_{\theta=\bar{x}} \right\}\hat{\sigma}^2 + \frac{1}{2}\left.\frac{\partial^3 L(\theta)}{\partial\theta^3}\right|_{\theta=\bar{x}}\hat{\sigma}^4. \tag{3.3.5}$$

Now recall from Example 3.2.1 that $\hat{\Sigma} = \hat{\sigma}^2 = \bar{x}/n$ and that $L(\theta) = -n\theta$

$+ (\log \theta)\Sigma x_j - \log(\Pi_j x_j!)$. Then

$$\left.\frac{\partial L(\theta)}{\partial \theta}\right|_{\theta=\bar{x}} = 0;$$

$$\left.\frac{\partial^2 L(\theta)}{\partial \theta^2}\right|_{\theta=\bar{x}} = -\frac{n}{\bar{x}}; \left.\frac{\partial^3 L(\theta)}{\partial \theta^2}\right|_{\theta=\bar{x}} = \frac{2n}{\bar{x}^2}.$$

Substituting into Eq. (3.3.5) gives

$$I(x_1,\ldots,x_n) \simeq \bar{x} + \left[\left.\frac{\partial \rho(\theta)}{\partial \theta}\right|_{\theta=\bar{x}}\right]\left(\frac{\bar{x}}{n}\right) + \left(\frac{n}{\bar{x}^2}\right)\left(\frac{\bar{x}^2}{n^2}\right)$$

$$= \bar{x} + \frac{1}{n}\left[1 + \bar{x}\left.\frac{\partial \rho(\theta)}{\partial \theta}\right|_{\theta=\bar{x}}\right].$$

Note also that since in this case, $I(x_1,\ldots,x_n) = E(\theta|x_1,\ldots,x_n)$, it follows that

$$E(\theta|x_1,\ldots,x_n) \simeq \bar{x} + \frac{1}{n}\left[1 + \bar{x}\left.\frac{\partial \rho(\theta)}{\partial \theta}\right|_{\theta=\bar{x}}\right]. \tag{3.3.6}$$

Since in this example, the prior distribution is log-normal, we have

$$\rho(\theta) = \log g(\theta) = -\log \sigma_0\sqrt{2\pi} - \log \theta - \frac{1}{2\sigma_0^2}(\log \theta - \mu)^2.$$

So

$$\left.\frac{\partial \rho(\theta)}{\partial \theta}\right|_{\theta=\bar{x}} = -\frac{1}{\bar{x}}\left[1 + \frac{\log \bar{x} - \mu}{\sigma_0^2}\right].$$

Substituting into eq. (3.3.6) gives

$$E(\theta|x_1,\ldots,x_n) \simeq \bar{x} + \frac{1}{n}\left[1 - \left(1 + \frac{\log \bar{x} - \mu}{\sigma_0^2}\right)\right],$$

or

$$E(\theta|x_1,\ldots,x_n) \simeq \bar{x} - \frac{1}{n}\left(\frac{\log \bar{x} - \mu}{\sigma_0^2}\right). \tag{3.3.7}$$

Thus, the parameters (μ, σ_0^2) of the prior distribution enter into the second-order approximation for the posterior mean. Note that the terms neglected in the approximation in Eq. (3.3.7) are $O(1/n^2)$.

The Tierney–Kadane Approximation

Another analytical approximation result that is quite useful for evaluation of Bayesian integrals is due to Tierney and Kadane (1984). It is given in Theorem 3.3.2. [The proof is based upon the steepest descent algorithm used for evaluation of Laplace integrals; see, e.g., Erdelyi (1956, p. 29).]

Theorem 3.3.2. For n sufficiently large, if the posterior distribution of $u(\theta)$ (given the data) is concentrated on the positive (or negative) half-line, and if $[L(\theta) + \rho(\theta)]$, defined in Eq. (3.3.1), concentrates around a unique maximum, under suitable regularity conditions, the ratio of integrals in Eq. (3.3.1) is given approximately by

$$I(x_1, \ldots, x_n) \simeq \frac{\sigma^*}{\sigma} \exp\{n[\mathscr{L}^*(\tilde{\theta}^*) - \mathscr{L}(\tilde{\theta})]\}, \tag{3.3.8}$$

where $n\mathscr{L}(\theta) \equiv L(\theta) + \rho(\theta)$, $n\mathscr{L}^*(\theta) = \log u(\theta) + L(\theta) + \rho(\theta)$, $\tilde{\theta}^*$ maximizes $\mathscr{L}^*(\theta)$, $\tilde{\theta}$ is the posterior mode and therefore maximizes $\mathscr{L}(\theta)$, and

$$\sigma^{-2} = -\left(\frac{\partial^2 \mathscr{L}(\theta)}{\partial \theta^2}\right)\bigg|_{\theta = \tilde{\theta}}, \qquad \sigma^{*-2} = -\left(\frac{\partial^2 \mathscr{L}^*(\theta)}{\partial \theta^2}\right)\bigg|_{\theta = \tilde{\theta}^*}.$$

Proof: See Tierney and Kadane (1984). □

REMARK 1: The result in Eq. (3.3.8) involves only second derivatives of $L(\theta)$, whereas the analogous result in Eq. (3.3.3) involves third derivatives of $L(\theta)$ as well.

REMARK 2: The terms omitted in the approximation in Eq. (3.3.8) are $O(1/n^2)$, as in the result in Eq. (3.3.3).

REMARK 3: While Theorem 3.3.1 holds for arbitrary functions $u(\theta)$, Theorem 3.3.2 holds only for $u(\theta) > 0$ [or $u(\theta) < 0$]. In some new work, it has been shown that the theorem can be modified so that it will hold for general $u(\theta)$. [See Tierney, Kass, and Kadane (1988).]

REMARK 4: It may be necessary to use numerical techniques to evaluate $\tilde{\theta}$ and $\tilde{\theta}^*$ in order to apply the theorem.

Example 3.3.2. As an illustration of Theorem 3.3.2, we consider the following example based upon the Poisson sampling distribution. We can consider the same context as in Example 3.2.1, where the likelihood function is

$$f(x_1,\dots,x_n|\theta) \propto e^{-n\theta}\theta^{\Sigma_1^n x_j}.$$

While we adopted a log-normal prior density in Example 3.2.1, here we adopt [along with Tierney and Kadane (1984)] a gamma density prior for θ, so that the posterior density is given by

$$h(\theta|x_1,\dots,x_n) \propto \theta^{\Sigma x_i+\alpha-1}e^{-(\beta+n)\theta},$$

where the prior density is

$$g(\theta) \propto \theta^{\alpha-1}e^{-\beta\theta}, \qquad \alpha > 0, \beta > 0.$$

Thus, if $\tilde{\alpha} \equiv \Sigma_1^n x_i + \alpha$ and $\tilde{\beta} \equiv \beta + n$, the posterior density becomes

$$h(\theta|x_1,\dots,x_n) \propto \theta^{\tilde{\alpha}-1}e^{-\tilde{\beta}\theta}.$$

The exact posterior mean is given by

$$E(\theta|x_1,\dots,x_n) = \frac{\tilde{\alpha}}{\tilde{\beta}}.$$

Suppose we wanted to approximate the posterior mean of θ by using Theorem 3.3.2. First we note from the structure of Eq. (3.3.1) that we only need to use the kernel of the likelihood function, as well as the kernel of the prior, rather than the entire normalized densities, since the proportionality constants of both cancel out in the ratio. Accordingly, set the log of the kernel of the likelihood equal to

$$L(\theta) = -n\theta + \Sigma x_i \log\theta,$$

and set the log of the kernel of the prior equal to

$$\rho(\theta) = (\alpha - 1)\log\theta - \beta\theta.$$

Thus,

$$\begin{aligned} n\mathscr{L}(\theta) &\equiv L(\theta) + \rho(\theta) \\ &= (\Sigma x_i + \alpha - 1)\log\theta - \theta(n + \beta) \\ &\equiv (\tilde{\alpha} - 1)\log\theta - \theta\tilde{\beta}. \end{aligned}$$

Similarly, taking $u(\theta) \equiv \theta$ in Theorem 3.3.2, take

$$\begin{aligned} n\mathscr{L}^*(\theta) &\equiv \log\theta + L(\theta) + \rho(\theta) \\ &= \log\theta + n\mathscr{L}(\theta) \\ &= \log\theta + (\tilde{\alpha} - 1)\log\theta - \theta\tilde{\beta} \\ &= \tilde{\alpha}\log\theta - \theta\tilde{\beta}. \end{aligned}$$

Next note that $n\mathscr{L}(\theta)$ is maximized at $\theta = \hat{\theta} = (\tilde{\alpha} - 1)/\tilde{\beta}$ and that $n\mathscr{L}^*(\theta)$ is maximized at $\theta = \hat{\theta}^* = \tilde{\alpha}/\tilde{\beta}$. Moreover, $\sigma^2 = (\tilde{\alpha} - 1)/\tilde{\beta}^2$, and $\sigma^{*2} = \tilde{\alpha}/\tilde{\beta}^2$. The posterior mean approximation is then found by substituting into Eq. (3.3.8). We find the approximation

$$\begin{aligned} \hat{E}(\theta | x_1, \ldots, x_n) &\equiv I(x_1, \ldots, x_n) \\ &\approx \frac{(\tilde{\alpha}^{1/2}/\tilde{\beta})(\tilde{\alpha}/\tilde{\beta})^{\tilde{\alpha}}\exp\{-\tilde{\alpha}\}}{[(\tilde{\alpha} - 1)^{1/2}/\tilde{\beta}][(\tilde{\alpha} - 1)/\tilde{\beta}]^{\tilde{\alpha}-1}\exp\{-(\tilde{\alpha} - 1)\}} \\ &= \left(\frac{\tilde{\alpha}}{\tilde{\beta}}\right)[\tilde{\alpha}/(\tilde{\alpha} - 1)]^{\tilde{\alpha}-1/2}e^{-1} \\ &= [E(\theta | x_1, \ldots, x_n)][\tilde{\alpha}/(\tilde{\alpha} - 1)]^{\tilde{\alpha}-1/2}e^{-1}. \end{aligned}$$

Note that we must have $\tilde{\alpha} > 1$ for the approximation to be applicable (otherwise we are estimating a positive quantity by one which is negative). Note that the relative error of the approximation depends only upon $\tilde{\alpha} = \Sigma x_i + \alpha$, and not upon sample size n. The table (Table 3.1) of relative errors in the approximation, given below, was presented by Tierney and Kadane (1984). Thus, as $\tilde{\alpha}$ increases from 2 to 10, the relative error (the approximate posterior mean, $\hat{E}$, compared with the true posterior mean, E) decreases from 4.05% to .097%.

REMARK: For Theorem 3.3.2 to be applicable we must have a unimodal $\mathscr{L}(\theta)$, as we had in this example. Unfortunately, however, some compli-

Table 3.1 Relative Errors in the Approximation

$\tilde{\alpha}$	2	3	4	6	8	10
$\hat{E}(\theta \mid x_1, \ldots, x_n)/E(\theta \mid x_1, \ldots, x_n)$ Relative error	1.0405	1.0138	1.0069	1.0028	1.0015	1.00097

cated posterior distributions are multimodal in all but extremely large samples, rendering the theorem inapplicable in such cases. Moreover, even when $\mathscr{L}(\theta)$ and $\mathscr{L}^*(\theta)$ are both unimodal, it may not be easy to find the modal values.

The Naylor–Smith Approximation

Analysis of Bayesian estimators in large samples, and comparison of them with maximum likelihood estimators, shows that the asymptotic normality assumption can be quite misleading. In many cases the asymptotic posterior distribution becomes symmetric only very slowly, so that a seemingly "large" sample size does not generate normality of the posterior. J. C. Naylor (1982) explored Gaussian quadrature methods of numerical integration to evaluate Bayesian integrals in several dimensions efficiently. Some of this work is described in Naylor and Smith (1982); see also Naylor and Smith (1983), Smith and Naylor (1984), and Smith et al. (1985). Classical quadrature techniques are presented, for example, in Davis and Rabinowitz (1967) and in Cohen et al. (1973). We summarize the Naylor–Smith approach below.

A classical result on numerical integration is that if $p_{2n-1}(x)$ denotes any polynomial of degree $(2n-1)$ in x, which is well defined in $[a, b]$, and $\{f_n(x)\}$ denotes a sequence of orthogonal polynomials in $[a, b]$ relative to a positive weighting function $w(x)$, such that $f_n(x)$ is a polynomial of degree n, then

$$\int_a^b w(x) p_{2n-1}(x)\, dx = \sum_{k=1}^{n} \alpha_k p_{2n-1}(x_k), \tag{3.3.9}$$

where the α_k's are coefficients, and the x_k's are the roots of $f_n(x) = 0$. The coefficients are given by

$$\alpha_k = \int_a^b \frac{w(x) f_n(x)}{(x - x_k) f_n'(x_k)}\, dx.$$

Proof: See Cohen et al. (1973, pp. 87 and 88). □

Various systems of orthogonal polynomials have been used. We select the Hermite class of orthogonal polynomials that are orthogonal relative to the weighting function

$$w(t) = \exp(-x^2),$$

Table 3.2 Zeros of Hermite Polynomials and Coefficients for Gauss–Hermite Quadrature

n	x_k	α_k
1	0	1.7724539
2	$\pm$.7071068	.8862269
3	0	1.1816359
	$\pm$1.2247449	.2954100
4	$\pm$.5246476	.8049141
	$\pm$1.6506801	.0813128
5	0	.9453087
	$\pm$.9585725	.3936193
	$\pm$2.0201829	.0199532

on the interval $[a, b] = [-\infty, \infty]$. In this case, Eq. (3.3.9) becomes the approximation

$$\int_{-\infty}^{+\infty} e^{-x^2} f(x) \approx \sum_{k=1}^{n} \alpha_k f(x_k),$$

where $x_1, \ldots, x_n$ are the roots of the Hermite polynomial equation $H_n(x) - 0$, and

$$\alpha_k = \frac{2^{n-1} n! \sqrt{\pi}}{n^2 [H_{n-1}(x_k)]^2}.$$

Integration methods of this (Hermite) form, based upon eq. (3.3.9), are called *Gaussian formulae*. If $f(t)$ is a polynomial of degree at most $(2n - 1)$, the approximation is exact. The first five zeros of $H_n(x)$ and the coefficient values of α_k are given in Table 3.2.

More generally, if $h(x)$ is a suitably regular function, and for some (μ, σ^2),

$$g(x) = \frac{h(x)}{(2\pi\sigma^2)^{1/2}} \exp\left\{\left(-\tfrac{1}{2}\right)\left(\frac{x-\mu}{\sigma}\right)^2\right\},$$

Naylor and Smith (1982) give the approximate relation

$$\int_{-\infty}^{\infty} g(x)\, dx \approx \sum_{k=1}^{n} m_k g(z_k). \tag{3.3.10}$$

where

$$m_k = \alpha_k \exp\left\{ \frac{x_k^2}{\sigma\sqrt{2}} \right\}, \qquad z_k = \left(\mu + \sigma x_k \sqrt{2}\right).$$

Tables of x_k, α_k [and $\alpha_k \exp(x_k^2)$] that are more extensive than Table 3.2 are available for $n = 1(1)20$ in Salzer et al. (1952). The error will be small if $h(z)$ is approximately a polynomial. The precision of the approximation in Eq. (3.3.10) depends upon the choices of μ and σ^2. A simple choice is to take (μ, σ^2) as the MLE's, but any prior guess could be used. Equation (3.3.10) could be used to evaluate posterior means or variances by taking $g(x)$ to be x^k times the posterior density, for $k = 1, 2$, and the approximation will be good as long as the posterior density is well approximated by the product of a normal density and a polynomial of degree at most $(2n - 3)$.

Readers interested in the extension of these results to problems with more than one parameter in the sampling density (several dimensions of numerical integration) should consult Naylor and Smith (1982).

3.3.2 High Dimension (by Monte Carlo Integration)

In Section 3.3.1 we discussed methods for approximating and numerically evaluating posterior densities and integrals for several parameters. Those methods are appropriate for several parameters but not for many parameters. When the order of integration (number of parameters) exceeds five or six, greater precision can be achieved with Monte Carlo integration and importance sampling. This approach is outlined below. See also Stewart (1979, 1983, 1984). Kloek and Van Dijk (1978) and Van Dijk and Kloek (1980, 1983, 1984) discuss this problem in an econometric context.

Suppose $\boldsymbol{\theta}: (k \times 1)$ denotes the (unobservable) parameter vector of a sampling distribution and we wish to evaluate integrals of the form

$$I(\mathbf{x}_1, \ldots, \mathbf{x}_n) = \int u(\boldsymbol{\theta}) h(\boldsymbol{\theta} | \mathbf{x}_1, \ldots, \mathbf{x}_n) \, d\boldsymbol{\theta}, \tag{3.3.11}$$

for some appropriate $u(\boldsymbol{\theta})$, as in Eq. (3.3.1). For motivation of the Monte Carlo procedure, rewrite this integral in the form

$$I(\mathcal{D}) = E_{\boldsymbol{\theta}}[u(\boldsymbol{\theta}) | \mathcal{D}], \tag{3.3.12}$$

where $\mathcal{D}$ denotes the data $(\mathbf{x}_1, \ldots, \mathbf{x}_n)$, and $E(\cdot)$ denotes the expectation taken over the k-dimensional $\boldsymbol{\theta}$-space.

Let $g^*(\boldsymbol{\theta})$ denote a "generating density" called the *importance function*. This k-dimensional density will be used to generate M points whose ordinates will be averaged to approximate $I(\mathscr{D})$. Let $\boldsymbol{\theta}_1, \ldots, \boldsymbol{\theta}_M$ be M points generated independently from $g^*(\boldsymbol{\theta})$. This is known as *importance sampling*. $g^*(\boldsymbol{\theta})$ is generally chosen to approximate the posterior density, but it is also chosen so that the $\boldsymbol{\theta}_m$'s can be easily generated. Equation (3.3.11) can be rewritten

$$I(\mathscr{D}) = \frac{\int u(\boldsymbol{\theta}) f(\mathscr{D}|\boldsymbol{\theta}) p(\boldsymbol{\theta})\, d\boldsymbol{\theta}}{\int f(\mathscr{D}|\boldsymbol{\theta}) p(\boldsymbol{\theta})\, d\boldsymbol{\theta}}$$

$$= \int u(\boldsymbol{\theta}) w(\boldsymbol{\theta})\, d\boldsymbol{\theta},$$

where

$$w(\boldsymbol{\theta}) = \frac{f(\mathscr{D}|\boldsymbol{\theta}) p(\boldsymbol{\theta})}{\int f(\mathscr{D}|\boldsymbol{\theta}) p(\boldsymbol{\theta})\, d\boldsymbol{\theta}}.$$

As an approximation to $I(\cdot)$ in Eq. (3.3.12), we take the weighted average of the $u(\boldsymbol{\theta}_m)$'s, namely,

$$\hat{I}(\mathscr{D}) = \sum_{m=1}^{M} \hat{w}(\boldsymbol{\theta}_m) u(\boldsymbol{\theta}_m), \tag{3.3.13}$$

where the weights $\hat{w}(\boldsymbol{\theta}_m)$ are

$$\hat{w}(\boldsymbol{\theta}_m) = \frac{f(\mathscr{D}|\boldsymbol{\theta}_m) p(\boldsymbol{\theta}_m)/g^*(\boldsymbol{\theta}_m)}{\sum_{m=1}^{M} [f(\mathscr{D}|\boldsymbol{\theta}_m) p(\boldsymbol{\theta}_m)/g^*(\boldsymbol{\theta}_m)]}.$$

Note that the weights sum to unity. We also note that because $f(\mathscr{D}|\boldsymbol{\theta}_m)$ and $p(\boldsymbol{\theta}_m)$ occur in ratio form, we will require only their kernels rather than their complete densities (since the normalizing constants of both cancel out in the ratio). Stewart (1983) points out that under easily attainable conditions, $\hat{I}(\mathscr{D})$ will converge almost surely to I, as M approaches infinity. The precision of the approximation will be heavily dependent upon the choice of $g^*(\boldsymbol{\theta})$, however. For additional discussion on how to choose the importance function, see Kloek and Van Dijk (1978, pp. 316 and 317).

3.4 SIMULATION OF BAYESIAN DISTRIBUTIONS

Sections 3.2 and 3.3 were concerned with approximations to posterior moments, integrals, densities, and so on. We have seen that exact distributional results are sometimes analytically intractable and that asymptotic approximations, when available, are sometimes not as accurate as we might prefer. Kass (1985) suggests a simple procedure for developing approximate posterior distributions in the case where we are interested in the posterior distributions of quantities based upon the covariance or correlation matrix. The approach is by simulation. It is explained below.

Suppose $\mathbf{Y}_1, \ldots, \mathbf{Y}_n$ are independent, identically distributed observations from $N(\boldsymbol{\mu}, \boldsymbol{\Sigma})$. We are interested in making posterior inferences about some function of $\boldsymbol{\Sigma}$, namely, $\phi(\boldsymbol{\Sigma})$—for example, the largest latent root of the correlation matrix, or of $\boldsymbol{\Sigma}$ (we might wish to make inferences about the largest latent root in principal components analysis, or factor analysis). Define the sample covariance matrix (unscaled)

$$\mathbf{V} \equiv \sum_{1}^{N} (\mathbf{Y}_i - \bar{\mathbf{Y}})(\mathbf{Y}_i - \bar{\mathbf{Y}})',$$

for $\bar{\mathbf{Y}} \equiv (1/N)\Sigma_1^N \mathbf{Y}_i$. The distribution of $\mathbf{V}$, for given $\boldsymbol{\Sigma}$, is Wishart (see Section 5.3). That is,

$$f(\mathbf{V}|\boldsymbol{\Sigma}) \propto \frac{1}{|\boldsymbol{\Sigma}|^{n/2}} \exp\left\{\left(-\tfrac{1}{2}\right)\operatorname{tr}\boldsymbol{\Sigma}^{-1}\mathbf{V}\right\},$$

for $n \equiv N - 1$, and $p \leq n$, $0 < \mathbf{V}$, $0 < \boldsymbol{\Sigma}$. (It may be noted that here we are concerned only with the dependency of the likelihood function upon $\boldsymbol{\Sigma}$. We have therefore suppressed a term of the density that depends upon $\mathbf{V}$ only, and incorporated it into the proportionality constant.)

Adopt the Jeffreys invariant vague prior density for $\boldsymbol{\Lambda} \equiv \boldsymbol{\Sigma}^{-1}$, namely,

$$p(\boldsymbol{\Lambda}) \propto \frac{1}{|\boldsymbol{\Lambda}|^{(p+1)/2}}.$$

By Bayes' theorem the posterior density for $\boldsymbol{\Lambda}$ is given by

$$p(\boldsymbol{\Lambda}|\mathbf{V}) \propto |\boldsymbol{\Lambda}|^{(n-p-1)/2} \exp\left\{\left(-\tfrac{1}{2}\right)\operatorname{tr}\boldsymbol{\Lambda}\mathbf{V}\right\}. \tag{3.4.1}$$

For a natural conjugate prior, suppose Λ is Wishart, so that for $0 < \Lambda$ we have

$$p(\Lambda) \propto |\Lambda|^{(m-p-1)/2} \exp\left\{\left(-\tfrac{1}{2}\right) \operatorname{tr} \mathbf{G}^{-1}\Lambda\right\},$$

$p \leq m$, $0 < \mathbf{G}$. Here, $(m, \mathbf{G})$ are hyperparameters, being assessed parameters of the prior distributions. By Bayes' theorem, it follows that

$$p(\Lambda|\mathbf{V}) \propto |\Lambda|^{(m+n-p-1)/2} \exp\left\{\left(-\tfrac{1}{2}\right) \operatorname{tr} \Lambda(\mathbf{V} + \mathbf{G}^{-1})\right\}. \quad (3.4.2)$$

Now suppose we want to study the posterior distribution of $\phi(\Sigma)$, some function of Σ. Most cases of ϕ of interest have posterior distributions that are analytically intractable. But we will approximate them by simulation. We can readily obtain a sample $\phi_1, \ldots, \phi_k$ from the posterior distribution of ϕ. The procedure is the following.

Generate a sample $\Lambda_1, \ldots, \Lambda_k$ from the appropriate posterior distribution of $\Lambda|\mathbf{V}$, using Eq. (3.4.1), Eq. (3.4.2), or some other appropriate posterior. For each Λ_j, calculate $\phi_j \equiv \phi(\Lambda_j^{-1})$, $j = 1, \ldots, k$. Using the sample $(\phi_1, \ldots, \phi_k)$ we may plot a histogram or a sample cdf, or we may fit an empirical density function. We may also calculate functions of the distribution of $\phi(\Sigma)$ such as the mean and standard deviation. The accuracy of this approach in providing the true posterior for $\phi(\Sigma)$ depends, of course, upon how large k is. Kass (1985) suggests that $k = 1000$ has proved to be more than adequate in all of the examples he has tried. Using crude (inefficient) simulation procedures, Kass found that a typical problem took less than 1 minute for simulation on a VAX/750.

The approach of this section is in the same spirit as that for the material presented in Section 2.10 with regard to the Bayesian bootstrap. The implication is that the simulation approach for studying posterior distributions is not limited to the assumption of underlying normality. In fact, other underlying assumptions for sampling distributions, as well as for accompanying priors, can be accommodated equally well. The simulation approach is a powerful one for studying the posterior distribution of functions of model parameters.

3.5 BAYESIAN COMPUTER PROGRAMS

3.5.1 Introduction

Provided here is a list of some of the Bayesian Computer Programs that are available today. This list was compiled, in part, from responses to a survey questionnaire that was sent to approximately 450 Bayesian statisticians and econometricians in January 1987. The information is up to date as of April 1987. Part of this list was compiled by Dr. Prem Goel (see also Goel, 1987) using the same format used in Press (1980a). Although Goel's listing is current, it does not include many of the special purpose programs found in the earlier listing. The list below is by no means complete, but it is intended to provide an initial source for potential users who might otherwise not have been aware of the existence of these programs. It should also be remembered that no package can remain up to date and that new programs and revised programs are continually being generated.

The program listing is divided into six parts. They are:

1. CADA, a general-purpose data analysis monitor
2. Normal linear regression, econometric models, and time-series analysis
3. Computation/approximation of posterior distributions
4. Elicitation of prior information
5. Reliability analysis
6. Miscellaneous

3.5.2 Program Summaries

1. CADA, a general-purpose data analysis monitor

Program name: CADA [Computer-Assisted Data Analysis Monitor (CADA Group)]

Function: This monitor provides a conversational language for Bayesian analysis. It has gone through several updatings. The most recent version (1983) includes substantial enhancements over the 1980 version. CADA is a hierarchically structured system with several component groups, namely: data management facility; simple parametric models; decision-theoretic models, full-rank ANOVA models; simultaneous estimation; full-rank MANOVA; exploratory data analysis; psychometric methods; probability distribution functions; and actuarial functions.

Input: Raw data to be entered in, on-line or data files to be loaded.

Output: Analysis for beta, two-parameter normal, and multinomial models using conjugate priors; assessment of conjugate priors and utility functions; full-rank Model I ANOVA and MANOVA for multifactor designs using conjugate for noninformative priors; elementary classical statistics;

graphical evaluations of various probability distributions; multiple linear regression analysis and simultaneous estimation of regression in *m*-groups.

Programming language: BASIC (compiler or interpreter required on the machine).

Machines: DEC-PDP-11(RSTS); DEC-VAX-11(VMS), PRIME, HP-3000. IBM PC version coming soon.

Documentation: Novick, M. L., Hamer, R. M., Libby, D. L., Chen, J. J., and Woodworth, G. G. (1983). *Manual for the Computer-Assisted Data Analysis (CADA) Monitor*. Iowa City, IA: The CADA Group, Inc.

Availability: The CADA Group, Inc., 306 Mullin Ave., Iowa City, IA 52240, telephone number (319) 351-7200.

Remarks: (i) The CADA system was developed at The Iowa State University under the direction of the late Professor Melvin R. Novick.
(ii) The software is fully supported and is available at a cost of $600 per copy.

2. *Normal linear regression, econometric models, and time-series analysis*

Program name: BRAP [Bayesian Regression Analysis Program (Abowd/Zellner)], version 2.0.

Function: This program provides a unified package for the Bayesian analyses of the normal linear multiple regression model (MRM) with multivariate normal errors under a noninformative prior, a *g*-prior, or a natural conjugate prior distribution. Some numerical integration capability via Simpson's rule and Monte Carlo importance sampling also provides a facility for analysis of nonstandard models. Both the prior and the posterior distributions of the regression coefficients can be analyzed. Plotting raw data and residuals, prior and posterior marginals, and bivariate contours for regression coefficients can also be done. The posterior distribution of linear functions of regression coefficients, the realized error terms, and predictive distribution of the dependent variables can also be obtained. Some transformations are already available, and IMSL® could be loaded for more transformations.

Input: The control cards are in JCL format. Data files can be easily loaded using a load command.

Output: Updates the prior parameters and plots marginal and bivariate contours of both the prior and the posterior distributions of the regression coefficients; posterior distribution of the realized errors; posterior distribution of linear functions of coefficients; standard posterior information; quantiles of posterior distribution can also be obtained via numerical integration routines.

Programming language: FORTRAN-IV

Machine: IBM-MVS (may need some modifications for newer IBM compilers).

Documentation: Abowd, J. M., Moulton, B. R., and Zellner, A. (1985). *The Bayesian Regression Analysis Package, BRAP User's Manual Version 2.0 of Dec. 1985.* H. G. B. Alexander Research Foundation, Graduate School of Business, University of Chicago.

Availability: The package is available from Professor Arnold Zellner, University of Chicago, Graduate School of Business, at a nominal cost.

Remarks: Other contributers to the development of BRAP include F. Finnegan, S. Grossman, C. Plosser, P. Rossi, A. Siow, J. Stafford, and W. Vandaele.

Program name: BAP [Bayesian Analysis Package (de Alba/Rocha)]

Function: This main program and subroutine package is an enhancement of BRAP in that is includes BRAP as well as subroutines for Bayesian disaggregation and constrainted forecasting.

Programming language: FORTRAN 77

Machines: IBM PC and PC compatibles

Availability: It is available from Professor Enrique de Alba, Instituto Technologico Autonomo De Mexico (ITAM), Rio Hondo, No. 1, Mexico, D.F. 01000, at a nominal cost of a diskette and mailing charges.

Program name: BATS [Bayesian Analysis of Time Series and Bayesian Forecasting]

Function: This package consists of a collection of functions that can be used to perform a variety of activities in data management, modeling, analysis, and forecasting and to produce many numeric and graphical displays, including pen-plotter output.

Programming language: APL*PLUS/PC

Machine: Personal microcomputers

Availability: June 1987, from the Bayesian Forecasting Group, Department of Statistics, University of Warwick, Coventry, CV4-7AL, England; Professors Mike West, Jeff Harrison, and Andy Pole.

Program name: SEARCH [Seeks Extreme and Average Regression Coefficient Hypothesis (Leamer/Leonard)]

Function: This is a user-oriented package for Bayesian inference and sensitivity analysis that pools prior beliefs about the regression coefficients with evidence embodied in a given data set. Prior beliefs are assumed to be equivalent to a previous, but possibly fictitious, data set. SEARCH offers a study of the sensitivity of the posterior estimates to changes in features of the prior beliefs expressed in terms of a fictitious data set.

Input: Formatted or free-format card-image files or on-line CRT input. Input files can be prepared on SAS®, BMDP®, TSP®, and SPSS®. SEARCH requires access to a double precision version of IMSL® library.

Output: Diagnostic messages for debugging syntax errors are available. The program reports the summary of prior and data information received and computes the approximate posterior mode for the regression coefficients. The prior beliefs are modeled as if they came from a normal population with a specified mean **r** and a covariance matrix **V**. It also reports the sensitivity of the modal estimate to changes in the prior location **r** and the prior covariance matrix **V** in the form of extreme bounds for any linear function of the parameters specified by the user.

Programming language: FORTRAN IV. The manual for Version 6 states that it is not completely available in FORTRAN source code. Several of the subroutines for performing high-precision arithmetic, which SEARCH calls for, are object code modules (written in IBM 370 machine code), and the bulk of the SEARCH is written in FORTRAN IV that is compiled at UCLA on the IBM FORTRAN IV G1 compiler (i.e., not necessarily ANSI standard FORTRAN IV).

Machine: IBM 370/3033

Documentation: Leamer, E. E., and Leonard, H. B. (1985). *User's Manual for SEARCH—A Software Package for Bayesian Inference and Sensitivity Analysis*, SEARCH Version 6, October 1985.

Availability: The program is available from Professor E. E. Leamer, Department of Economics, UCLA, 405 Hilgard Ave., Los Angeles, CA 90024, telephone number (213) 825-1011, at $100 per copy on an IBM OS standard label nine-track 1600 BPI tape containing four card-image files. Cards or a 6250 BPI tape can be made available on special request.

Remarks: (i) The program was developed by Edward E. Leamer and Herman B. Leonard. This version was programmed by Arvin Stidick, and the MANUAL was extensively edited and largely rewritten by Thomas E. Wolff.

(ii) The Version 6 differs from Version 5 in its efficiency of computation and economy of input and output.
(iii) A latest example of how SEARCH can be used is given in Leamer, E. E., and Leonard, H. B. (1983). Reporting the fragility of Regression Estimates, *The Review of Economics and Statistics*.

Program name: MICRO EBA [Microcomputer version of SEARCH (Fowles)]

Function: The main program is the microcomputer version of Leamer and Leonard's program SEARCH described above.

Programming language: GAUSS

Machine: Any personal computer running GAUSS software package Version 1.46 or higher.

Availability: The program is available free of charge from Professor Richard Fowles, Department of Economics, Rutgers University, Newark, NJ 07102.

Program name: BRP [Bayesian Regression Program (Bauwens)]

Function: The main program executes computations necessary for Bayesian regression analysis for various standard econometric models. The prior beliefs are modeled as poly-**t** densities.

Input: Raw data as card-image files. Input data are echoed as output.

Output: Posterior parameters and marginals of regression coefficients and precision and standard deviations; classical regression analysis, posterior residuals, and predictive density function of the dependent variable; conditional posterior with given precision, conditional posteriors of some regression coefficients given some others, marginalized over the precision.

Programming language: FORTRAN 77

Machine: IBM 370/158 at the University of Louvain. It is portable according to Dr. Bauwens.

Documentation: Bauwens, L., and Tompa, H. (1977). *Bayesian Regression Program* (*BRP*), CORE User's Manual Set #A-5. See also Tompa, H. (1977). *Poly-t Distributions* (*PTD*), CORE User's Manual Set #C-9.

Availability: It is available from Professor Luc Bauwens, CORE, 34 Voie Du Roman Paays, B-1348 Louvain-La-Neuve, Belgium at a cost of 5000 Belgium francs.

Remarks: (i) BRP calls another program PTD to evaluate poly-**t** densities.
(ii) These programs have been developed by H. Tompa under the

guidance of Professor Jacques Dréze and Jean-Francois Richard and with assistance from Luc Bauwens, Jean-Paul Bulteau, and Philippe Gille.

Program name: Fully Bayesian Analysis of ARMA Time-Series Models (Monahan)

Function: A collection of main program and subroutines carries out the Bayesian analysis for ARMA time-series models using natural conjugate priors.

Input: Information not yet available.

Output: Programs compute the posterior and predictive distribution of parameters for a given set of ARMA models using the natural conjugate prior. The graphical displays can be obtained via SAS/GRAPH.

Programming language: FORTRAN 66

Machine: Portable

Documentation: Monahan, J. (1980). *A Structured Bayesian Approach to ARMA time series models, I, II, III*, Technical Reports, Department of Statistics, North Carolina State University, Raleigh, NC.

Availability: The programs package is available on tape from Professor John Monahan, Department of Statistics, North Carolina State University, P.O. Box 8203, Raleigh, NC 27695 at a nominal charge.

Program name: Sampling the Future (Thompson)

Function: This program simulates the predictive distribution of a set of future observations via Monte Carlo methods.

Output: The main program and several subroutines provide a Monte Carlo histogram for the predictive distribution of a future observation or a scattergram of samples from the predictive distribution of a pair of future observations. The model may contain as many as 10 ARMA parameters in up to three AR factors and up to three MA factors. Thus, multiplicative seasonal factors and the difference factors may be used in the model. The estimation step allows either a diffuse or a conjugate normal/gamma prior distribution.

Programming language: FORTRAN 77 ANSI standard.

Machine: The program should run on any machine with a standard FORTRAN 77 compiler and an IMSL® library. Future extensions will require a graphics terminal. The program will run on a PC with a math co-processor, but an AT type machine with a hard disk is recommended for realistic usage.

Availability: The package is available on a diskette, for a nominal charge of $10, from Professor Patrick Thompson, Faculty of Management Sciences, The Ohio State University, 1775 S. College Road, Columbus, OH 43210.

Remarks: Future plans include (1) a graphic display of predictive distributions and (2) the addition of the algorithm for prediction from a set of ARMA models given in Monahan (1983).

Program name: Bayes & Empirical Bayes Shrinkage Estimation of Regression Coefficients (Nebebe)

Function: The program computes Bayes and empirical Bayes estimates for a multiple normal linear regression model in which the prior for the regression coefficients and the precision is modeled as a hierarchical normal with mean μ and precision τ^2 and the hyperparameters are assumed to have various diffuse distributions.

Programming language: FORTRAN, requires access to NAG Library.

Documentation: No separate documentation is available. The details are given in Nebebe, F. (1984). Ph.D. thesis, Department of Mathematics and Statistics, Queen's University, Kingston, Canada.

Availability: The program is available from Professor F. Nebebe, Department of Decision Sciences and MIS, Concordia University, 1455 De Maisonnevue Blvd. West, Montreal, Quebec H3G 1M8, Canada.

Remarks: This program does not appear to provide any capability that is not available in BRAP, SEARCH, or BAP.

Program name: SHAZAM [General Econometrics Program (White)]

Function: The program provides a portable FORTRAN program for general econometric modeling on a PC for $250, or on a main frame for $500–$900.

Availability: Available from Professor Kenneth J. White, Economics Department, University of British Columbia, Vancouver, B.C., Canada.

Program name: BTS [Bayesian Time Series (Carlin/Dempster)]

Function: The program package carries out computations for Bayesian estimation of unobserved components ("seasonal"/"nonseasonal") in monthly time series under a class of Gaussian mixed models. It uses likelihood-based methods for estimation of model parameters.

Output: The program provides posterior estimates of model parameters. A nonportable version for the Apollo DN600 workstation has many graphical capabilities.

Programming language: FORTRAN 77 (Standard ANSI)

Documentation: Description of the program is available in Carlin, J. B. (1987). Ph.D. Thesis, Department of Statistics, Harvard University.

Availability: Available free of charge from Professor A. P. Dempster, Department of Statistics, Harvard University, Science Center, 1 Oxford Street, Cambridge, MA 02138.

Program name: PROC SEQ [Sequential Scoring Algorithm (Blattenberger)]

Function: The function performs iterative computation of the forecasting distribution for the dependent variable of a normal linear model with a normal-gamma prior distribution or optional *g*-priors. Scores for five different scoring rules are also computed.

Programming language: STAT 80 Procedure, currently being converted to SAS® PROC MATRIX.

Availability: Available free of charge from Professor Gail Blattenberger, Department of Economics, University of Utah, Salt Lake City, UT, 84112.

Program name: MAXENT [Data Analysis by Maximum Entropy Principle Version 1.17 (Jaynes)]

Function: This beta test version of MAXENT provides fitting of an incompletely specified linear model of the form $\mathbf{Y} = \mathbf{XF}$, where the data vector is $\mathbf{Y}$, the "smearing matrix" $\mathbf{X}$ is known but does not have full rank, and the elements of the vector $\mathbf{F}$ are non-negative and add to 1. The Maximum Entropy Principle (see Jaynes, 1983) finds the solution to the above equation which maximizes the entropy of the probability distribution $\mathbf{F}$.

Input: The program is interactive. One needs to decide the accuracy level for satisfying all the constraints.

Output: The optimal solution is obtained in an iterative mode. The output for each iteration can be printed.

Programming language: BASIC

Machines: IBM PC and compatibles. An ASCII source code file is also on the diskette for transporting the program to other microcomputers.

Documentation: There is a help file as well as a manual on the diskette.

Availability: Available free of charge from Professor Ed T. Jaynes, Department of Physics, Washington University, St. Louis, MO 63130. Individuals sending comments and user experience to Professor Jaynes will receive the Version 2.0 free.

The programs briefly discussed below have been written for applications of linear models to specific problems.

Program name: RECONDA (Braithwait, Steven)

Function: This program incorporates engineering prior estimates of appliance level electricity consumption into a statistical analysis of household hourly consumption via a hierarchical linear model.

Programming language: C

Machine: IBM PC and PC compatibles

Availability: The program will be distributed free of charge by EPRI, P.O. Box 10412, Palo Alto, CA 94303 to EPRI member utilities as well as to government and academic institutions.

Program name: Statistical Cost Allocation (Wright, Roger)

Function: Implements the indirect cost allocation methodology based on a multiple linear model.

Programming language: FORTRAN 77 (Standard ANSI)

Documentation: The program description and listing are given in Wright, R., and Oberg, K. (1983). *The 1979–80 University of Michigan Heating Plant and Utilities Cost Allocation Study*, Working Paper #352, Graduate School of Business Administration, The University of Michigan.

Availability: Available free of charge from Professor Roger Wright, Graduate School of Business Administration, The University of Michigan, Ann Arbor, MI 48109.

3. *Computation / approximation of posterior distributions*

Program names: Bayes Four and *gr* (Smith, A. F. M.)

Function: The Bayes Four system consists of a library of subroutines which is primarily intended for numerical computation of multiple integrals in interactive mode. The posterior distribution's features are evaluated for up to six parameters using numerical integration procedures and for up to 20 parameters using Monte Carlo integration. The *gr* library consists of subroutines for an interactive color graphics system which can be used to reconstruct and display output of the BAYES FOUR system; for reference, see Smith et al. (1985).

Input: Solving an inference problem requires writing an additional program code for a specific problem which can call BAYES FOUR and *gr* subroutines.

Output: The posterior moments and marginals can be evaluated by calling these menu-driven subroutines. The *gr* package can be used to provide graphical displays of the univariate and bivariate marginal posterior densities and predictive densities from outputs of BAYES FOUR.

Programming language: BAYES FOUR in FORTRAN 77; *gr* in 68000 assembler, C, and FORTRAN 77.

Machine: BAYES FOUR is portable. However, *gr* has not been configured for any standard graphics system or workstation yet.

Documentation: Naylor, J. C., and Shaw, J. E. H. (1985), *BAYES FOUR—User Guide*; Naylor, J. C., and Shaw, J. E. H. (1985), *BAYES FOUR—Implementation Guide*; and Shaw, J. E. H. (1985). *gr User Guide*. All of these are technical reports from the Nottingham Statistics Group, Department of Mathematics, University of Nottingham.

Availability: These systems may be obtained from Professor Adrian Smith, Department of Mathematics, University of Nottingham, Nottingham, U.K.

Remarks: The Nottingham Statistics Group is actively involved in developing numerical integration systems for implementing Bayesian methodology. Therefore, some enhanced versions of these subroutine packages may be available soon.

Program name: Simple Importance Sampling [Computation of Posterior Moments and Densities via Monte Carlo Integration (Van Dijk)]

Function: The program approximates multiple integrals that arise in the posterior moments and marginal densities of parameters of interest in econometric and statistical modeling, via the Monte Carlo integration method known as *importance sampling* (see Sect. 3.3).

Programming language: FORTRAN 77

Documentation: The algorithm, the program listing, and some examples are given in Van Dijk, H. K., Hop, J. P., and Louter, A. S. (1986). *An Algorithm for the Computation of Posterior Moments and Densities Using Simple Importance Sampling*, Econometric Institute Report 8625/A, Erasmus University, Rotterdam.

Availability: The program is available from Professor Herman K. Van Dijk, Econometric Institute, Erasmus University Rotterdam, P.O. Box 1738-3000 Dr., Rotterdam, The Netherlands.

Remarks: The program provides some interesting methods for constructing importance sampling densities which are more flexible than the multivariate Student's t density. There is also a PC-AT version of a Monte Carlo integration program which uses these more flexible methods. It is available on diskettes from Professor John Geweke, Institute of Statistics and Decision Sciences, Duke University, Durham, NC 27706.

Program name: BAYES 3/3D [Multiparameter Univariate Bayesian Analysis using Monte Carlo Integration (Stewart)]

Function: Bayesian inference for a univariate response variable is carried out by using Monte Carlo integration. Up to nine-parameter flexibility is allowed. The program can handle the usual random sampling data, interval data, censored data, and binomial data at different stresses or times.

Input: Data and control cards as card-image files.

Output: Displays posterior means and posterior percentile curves for CDF's, hazard rate functions, or probability of failure (response) versus stress (dose) or time; see Stewart (1979, 1983, 1984).

Programming language: FORTRAN 77

Machine: A graphics terminal is highly desirable but not absolutely necessary. Need DISPLA graphics software. GKS and DI-3000 versions are being written.

Documentation: Stewart, L. (1987). *User's Manual for BAYES 3/3D*, a program for multiparameter univariate Bayesian analysis using Monte Carlo integration.

Availability: The program was developed under various federal contracts at Lockheed Palo Alto research Laboratory, Palo Alto, CA 94304. Dr. Leland Stewart will provide the tapes when they can be made available.

Program name: LINDLEY.BAS (Sloan)

Function: This BASIC subroutine performs algebraic manipulation and constructs the expanded formula for use in approximating the ratio of two integrals, required in the evaluations of the posterior distribution's features, as discussed in Lindley (1980), and in this chapter.

Input: The program prompts for the number of parameters to be estimated.

Output: The printout gives the complete algebraic equation needed to approximate the ratio of integrals.

Programming language: MS BASIC

Machine: IBM PC or compatibles. Special printing customized for EPSON series of printers.

Availability: Available free of charge from Professor Jeff A. Sloan, Department of Statistics, University of Manitoba, Winnipeg, Manitoba, Canada R3T 2N2.

Program name: SBAYES (Tierney)

Function: The system consists of *S*-functions to compute approximations of posterior means, variances, and marginal densities, see Tierney, Kass, and Kadane (1986).

Programming language: FORTRAN 77 and C. Requires access to the S package for implementation.

Availability: Available free of charge from Professor Luke Tierney, School of Statistics, University of Minnesota, Minneapolis, MN 55113.

4. Elicitation of prior information

Program name: BAYES (Schervish)

Function: This program elicits priors and finds posterior and predictive distributions for samples from normal or binomial data with natural conjugate priors or mixed conjugate plus point mass priors. It also handles flat priors over bounded regions for normal data.

Programming language: FORTRAN IV, requires access to IMSL.

Machine: DEC-2060. Graphics are good for GIGI terminals only.

Availability: The program is not yet ready for distribution. Available on request from Professor Mark Schervish, Department of Statistics, Carnegie-Mellon University, Pittsburgh, PA 15213.

Program name: [B/D] [Beliefs Adjusted by Data (Goldstein/Wooff)]

Function: The program is in final development stage, and it provides an interactive, interpretive subjectivist analysis of general (partially specified, exchangeable) beliefs.

Output: The program output provides summaries of how and why beliefs are (i) expected to change and (ii) actually change; it also provides system diagnostics based on comparison of (i) and (ii).

Programming language: PASCAL

Availability: Will be available at cost of mailing and manual production from Professor Michael Goldstein, Department of Statistics, University of Hull, Cottingham Road, Hull, U.K.

5. *Reliability analysis*

Program name: BASS [Bayesian Analysis for Series Systems (Martz)]

Function: This program performs a Bayesian reliability analysis of series systems of independent binomial subsystems and components for either prior or test data at the component, subsystem, and overall system level. It uses a beta prior for the survival probabilities.

Programming language: FORTRAN 77

Machine: Portable but requires DISPLA software package for graphics.

Availability: Free of charge from Dr. Harry F. Martz, Group S-1, MS F600, Los Alamos National Laboratory, Los Alamos, NM 87545.

Program name: BURD [Bayesian Updating of Reliability Data (Martz)]

Function: The program performs Bayesian updating of binomial and Poisson likelihoods with a natural conjugate prior or a log-normal prior for the parameter. The updating for a log-normal prior is done via Monte Carlo integration. These models are used in the nuclear industry. The program belongs to Babcox and Wilcox, Inc.

Documentation: Ahmed, S., Metcalf, D. R., Clark, R. E. and Jacobsen, J. A. (1981). BURD—*A Computer Program for Bayesian Updating of Reliability Data*, NPGD-TM-582, Babcox and Wilcox, Inc., Lynchburg, VA.

Program name: IPRA [An Interactive PC-based Procedure for Reliability Assessment (Singpurwalla)]

Function: A menu-driven program performs a prior assessment based on expert opinion or informed judgment for Weibull-distributed life-length data and the posterior analysis in a highly interactive manner. It also allows the incorporation of the analyst's opinion on the expertise of the experts.

Input: On-line data entry of use of menu option to store data in a file for a later use in the analysis.

Output: The program computes the marginal and joint posterior densities of the Weibull parameters. The prior and posterior reliability functions for a specified time interval, as well as distributions of reliability for specified mission times, can be computed. These quantities can be displayed in a tabular or 2-d/3-d graphics form or can be saved on disk.

Programming language: IBM BASIC

Machines: IBM PC or compatibles with math co-processor and graphics board.

Documentation: Aboura, K. N., and Soyer, R. (1986). *A User's Manual for an Interactive PC-Based Procedure for Reliability Assessment*, Technical Report GWU/IRRA/Serial TR-86-14, George Washington University, Washington, D.C.

Availability: The program diskette and user's manual are available from Professor Nozer Singpurwalla, Department of Operations Research, George Washington University, Washington, D.C. 20052 at a nominal charge.

Program name: IPND [An Interactive PC-Based System for Predicting the Number of Defects due to Fatigue in Railroad Tracks (Singpurwalla)]

Function: A menu-driven program performs a Bayesian analysis of a non-homogeneous Poisson process with a Weibull intensity function in which the assessment of the prior information about the parameters is induced via an engineering model based on *S–N* curves. The procedure is applied to the prediction of the number of defects due to fatigue in railroad tracks.

Input: On-line data entry or use of menu option to store data in a file for a later use in the analysis.

Output: The program computes the marginal and joint posterior densities of the parameters in the Weibull intensity function. The prior and posterior distribution of the number of defects due to fatigue over a time period is also computed. These quantities can be displayed in a tabular or 2-d/3-d graphics form or can be saved on disk.

Programming language: IBM BASIC

Machines: IBM PC or compatibles with math co-processor and graphics board.

Documentation: Choksy, M., and Daryanani, S. (1987). *An Interactive PC-Based System for Predicting the Number of Defects due to Fatigue in Railroad Tracks: User's Manual*, Technical Report GWU/IRRA/Serial TR-87-3, George Washington University, Washington, D.C.

Availability: The program diskette and user's manual are available from Professor Nozer Singpurwalla, Department of Operations Research, George Washington University, Washington, D.C. 20052 at a nominal charge.

Remarks: This procedure, along with the program, has been adopted by the Association of American Railroads for the analysis of fatigue defects data in railroad tracks. This is an indication that availability of appropriate software would lead to a widespread use of Bayesian methodology.

Program name: PREDSIM [Prediction and Simulation for Mixtures of Exponentials (Sloan)]

Function: This program performs a Monte Carlo simulation of sampling from a mixture-of-exponentials model using a method proposed by Marsaglia, and computes Bayes estimates of the systematic parameters and reliability function and predictive intervals for future observations.

Programming language: PL/I

Machine: Portable but requires access to IMSL.

Availability: Available free of charge from Professor Jeff A. Sloan, Department of Statistics, University of Manitoba, Winnipeg, Manitoba, Canada R3T 2N2.

6. *Miscellaneous*

Program name: DISCBDIF (Stroud)

Function: This SAS® program classifies an input record into one of the two normal populations, based on training samples for each one. It uses either Geisser's discrimination procedure or a semidiffuse limit of conjugate priors.

Programming language: Requires access to SAS® package and SAS® PROC MATRIX.

Availability: Available free of charge from Professor Thomas W. F. Stroud, Department of Mathematics and Statistics, Queen's University, Kingston, Ontario, Canada K7L 3N6.

3.6 SUMMARY

In this chapter we have focused upon the importance of approximations, numerical methods, and computer-program-assisted solutions to problems of Bayesian inference. We presented the large-sample normal approximation to posterior distributions (the effect of the prior disappears in very large samples). Then we showed that the large-sample distribution could be improved with one that took the prior into account as well. We presented approximations to Bayesian integrals depending upon (1) numerical integration (in a few dimensions) and (2) Monte Carlo importance sampling (for many-dimensional integrals). We showed how to study posterior distributions by means of simulations, and we provided a catalog of some Bayesian computer programs currently available.

EXERCISES

3.1 Let $X_1, \ldots, X_N$ denote i.i.d. observations from the distribution

$$f(x \mid \beta) = \beta e^{-\beta x}, \qquad x > 0, \beta > 0.$$

Adopt a natural conjugate prior distribution for β and find:

(a) the posterior density kernel for β given $x_1, \ldots, x_N$;

(b) the large-sample posterior normal distribution for β, given $x_1, \ldots, x_N$.

3.2 Use the Lindley approximation of Theorem 3.3.1 to approximate the posterior mean in Exercise 3.1, part (a).

3.3 Give the Tierney–Kadane approximation for the posterior variance in Exercise 3.1, part (a).

3.4 Explain the method of Gauss–Hermite quadrature, and explain how you would use it to evaluate

$$I = \int_0^1 \frac{e^{-x}}{(1+x)^{10}}\, dx,$$

using a three-point quadrature grid. (*Hint*: take $n = 3$ in Table 3.2.)

3.5 Explain what is meant by "importance sampling." When would you use it?

3.6 Explain how you would use simulation of the posterior distribution to evaluate the posterior distribution of σ^3, where

$$y_i | x_i = a + bx_i + \varepsilon_i, \qquad i = 1, \ldots, n,$$

$\varepsilon_i \sim N(0, \sigma^2)$, the ε_i are uncorrelated, (a, b) are unknown coefficients, and the prior distribution is

$$g(a, b, \sigma^2) = g_1(a, b) g_2(\sigma^2),$$

$$g_1(a, b) \propto \text{constant},$$

$$g_2(\sigma^2) \propto \frac{1}{\sigma^2}.$$

3.7 Which computer program would you use if you intended to carry out a Bayesian analysis of the coefficients in a univariate multiple regression? Why?

CHAPTER IV

Assessment of Multivariate Prior Distributions: Assessing the Probability of Nuclear War

4.1 INTRODUCTION

You are a decision maker who would like to be helped in making a meaningful judgment.

There are several correlated observables as well as several correlated unobservables. How might we generate multivariate subjective probability distributions for the unobservables? One answer is to use the combined judgments of an informed group to assess a prior distribution for the unobservables and then to use Bayes' theorem to make posterior inferences conditional on the observables.

In this chapter we do the following:

1. We propose a model for grouping opinions.
2. We describe elicitation procedures, based upon the model, for assessing a multivariate prior density.
3. We show how to merge the points assessed into a density.
4. We report the results of an empirical application involving assessing a prior probability density for the event of nuclear war.
5. Finally, we summarize the biases introduced into the subjective probability assessments by psychological factors.

A more detailed discussion of this problem, including other approaches, can be found in Press (1985a).

4.2 MULTIVARIATE SUBJECTIVE ASSESSMENT

People cannot easily think in many dimensions simultaneously. For example, they are overwhelmed by being asked for their subjective probability that a random variable X_1 lies in a given interval (a_1, b_1), and simultaneously, a second random variable X_2 lies in another interval (a_2, b_2). It is much harder, of course, if we ask for their joint probability for p variables, and require:

$$a_1 \leq X_1 \leq b_1, \qquad a_2 \leq X_2 \leq b_2, \ldots, a_p \leq X_p \leq b_p.$$

One of the most important problems of Bayesian inference is how to improve our methods of assessing prior distributions, especially multivariate prior distributions. While the approach we propose here will not be applicable in all situations, it should provide a useful solution in many problems. The approach we propose here is applicable to assessment of both univariate and multivariate prior distributions.

4.3 OVERVIEW

Our proposal involves the simple idea that instead of asking a single individual to generate a complete p-dimensional assessed distribution, we assume that the individual is willing to substitute for his/her prior the distribution generated by the opinions of N other individuals, treating each opinion as a sample point from some common underlying distribution. This is just a formalization of the old idea of using a consensus view of informed individuals to express your judgment about an important proposition.

In the multivariate case, every opinion of a subject in the group is a p-dimensional vector. Thus, each subject can be asked for his/her p responses (correlated) to each of p questions. (A questionnaire with p items could be used.)

4.4 MODEL

Suppose that $\boldsymbol{\theta}: (p \times 1)$ is a continuous, unobservable, random vector denoting the mean of some p-dimensional sampling distribution with density $f(\mathbf{z} | \boldsymbol{\theta})$. You would like to assess your prior probability density for $\boldsymbol{\theta}$, and you would like your prior density to reflect the combined judgments of some informed population. We assume the members of the informed population are "experts," in that they possess contextual knowlege. We take a sample from the population of experts and ask each subject for a "best guess" opinion about $\boldsymbol{\theta}$. Let $\boldsymbol{\theta}_j$: $(p \times 1)$ denote the vector of opinions about $\boldsymbol{\theta}$ given by the jth subject, for $j = 1, \ldots, N$.

Note: We must be careful to ask each expert for his/her opinion vector privately and independently of all other experts. Then we can guarantee that the $\boldsymbol{\theta}_j$'s are mutually independent (but the assessments within a $\boldsymbol{\theta}_j$ vector are all correlated, because they are responses to related questions by the same individual).

4.5 MULTIVARIATE DENSITY ASSESSMENT

Assume that N assessments of the expert panel $\boldsymbol{\theta}_1, \ldots, \boldsymbol{\theta}_N$, all $(p \times 1)$, are independent and identically distributed, all with density $g(\boldsymbol{\theta})$. We adopt the (kernel density estimation) approach of Parzen (1962) and Cacoullos (1966) to density assessment [for discussion of general methodology of this subject, see, e.g., Taipia and Thompson (1982) or Silverman (1986); for discussion relevant to this application, see Press (1983) and the references therein]. We permit the density to assume the form

$$\underset{(p\times 1)}{g_N(\boldsymbol{\theta})} = \frac{1}{N}\sum_{j=1}^{N} \underset{(p\times 1)}{\phi_j(\boldsymbol{\theta})},$$

where

$$\phi_j(\boldsymbol{\theta}) = \frac{1}{\delta^p(N)} K\left(\frac{\boldsymbol{\theta} - \boldsymbol{\theta}_j}{\delta(N)}\right).$$

$K(\boldsymbol{\theta})$ is a kernel chosen to satisfy suitable regularity conditions, and $\delta(N)$ is a sequence of positive constants satisfying

$$(1) \quad \lim_{N\to\infty} \delta(N) = 0$$

and

$$(2) \quad \lim_{N\to\infty} N\delta^p(N) = \infty.$$

Normal Kernel

For simplicity, we adopt the normal density kernel (the result is quite insensitive to the parametric form of the kernel assumed),

$$K(\boldsymbol{\theta}) = \frac{1}{(2\pi)^{p/2}} e^{-\boldsymbol{\theta}'\boldsymbol{\theta}/2},$$

and we take

$$\delta(N) = \frac{C}{N^{1/(p+1)}},$$

where C is any preassigned constant (fixed in any given sample to smooth the density). Thus, if $\mathbf{X}_j$ has density $\phi_j(\mathbf{x})$, we have

$$\underset{(p\times 1)}{\mathbf{X}_j} \sim N\big(\boldsymbol{\theta}_j, \delta^2(N)\mathbf{I}_p\big).$$

Moreover,

$$g_N(\boldsymbol{\theta}) = \frac{1}{C^p N^{1/(p+1)}} \sum_{j=1}^{N} \frac{\exp\{[-N^{2/(p+1)}/2C^2](\boldsymbol{\theta} - \boldsymbol{\theta}_j)'(\boldsymbol{\theta} - \boldsymbol{\theta}_j)\}}{(2\pi)^{p/2}}.$$

Also, $g_N(\boldsymbol{\theta})$ is asymptotically unbiased, and

$$\lim_{N\to\infty} E[g_N(\boldsymbol{\theta}) - g(\boldsymbol{\theta})]^2 = 0,$$

at every point of continuity $\boldsymbol{\theta}$ of $g(\boldsymbol{\theta})$.

4.6 SUMMARY OF GROUP ASSESSMENT APPROACH

1. *Consensus of a Group of Experts Is Not Required, as in Some Group Procedures.* It is the diversity of opinions of experts that is interesting. When a group of experts is given some battery of questions independently, their degree of disagreement reflects the amount of inherent uncertainty there is in the underlying issue.

2. *No Need for an Assumed Functional Form of the Prior.* We don't use natural conjugates, or any other family of priors, and then assess hyperparameters. We merely let the actual empirical distribution of expert opinion determine the prior.

3. *Assessment Ease.* It is not necessary for subjects to think about the likelihood of certain events occurring simultaneously, in order to assess a p-vector. We need only to administer a battery of p-questions to N people, and each person's p-responses are correlated. Subjects need only to think about one-dimensional (marginal) events. It becomes as easy to assess a p-dimensional prior as it is to assess a one-dimensional prior.

4. *Additional Fractiles.* It is usually of interest to assess several fractiles of $g(\theta)$ from each subject (such as medians and quartile points, or means and variances, etc.). This second moment information reflects how certain each respondent is about his/her belief regarding θ. Such information could be used in a variety of ways. The "best guess" assessments of θ could be weighted by precisions (reciprocal variances) provided by the respondents; high-variance assessments of θ could be ignored; and so on. For an illustration of how we have used 90% assessed fractiles in our empirical application to nuclear war in Section 4.7, see Figure 4.4 (we used 5% assessments in each tail).

5. *Convergence to "Truth".* If there is indeed "truth," in some absolute sense, increasing the size of the group will not necessarily cause convergence to truth. In groups whose opinions follow substantially different distributions, such convergence is unlikely. In groups with identically distributed opinions, convergence is guaranteed. In group situations in between, it is hard to say.

4.7 EMPIRICAL APPLICATION: NUCLEAR WAR

It was of interest in 1980 to know how experts in the United States felt about the chances of nuclear war between the United States and the Soviet Union during the 1980s. A list was prepared containing the names of experts in strategic policy who were employed at The Rand Corporation or who were members of the California Arms Control Seminar. A random sample of 149 experts was selected; they were asked to complete written questionnaires. Each subject was asked in early 1981:

> Please express your subjective assessment that the United States or the Soviet Union will detonate a nuclear weapon in the other's homeland during the year 1981.

Denote the event of this occurence by E, and its probability by $P\{E\}$. Subjects were asked to respond on an *ordinal scale* (by circling one "+" point on the scale below).

Ordinal Scale

+ The probability is greater than the probability of getting a "head" by flipping a fair coin once. (The chance of getting a "head" is 0.5.)
+ The probability is less than the probability of getting a "head" by flipping a fair coin once but is greater than the probability that a

randomly selected state in the United States would be the state in which you live. (The actual chance that a randomly selected state would be yours is 2%.)

\+ The probability is less than the probability that a randomly selected state would be the one in which you live but is greater than the probability of throwing boxcars (double sixes) in two successive throws of the dice. (The actual chance of throwing double boxcars is 1 in 1,296.)

\+ The probability is less than the probability of throwing double boxcars in two successive throws of the dice but is greater than the probability that a randomly selected person in Los Angeles would be you if you lived in Los Angeles. (There are about 2 million people living in Los Angeles.)

\+ The probability is less than the chance that a randomly selected person in Los Angeles would be you if you lived in Los Angeles but is greater than the probability that a randomly chosen American will be you. (There are about 220 million Americans.)

\+ The probability is less than the probability that a randomly chosen American will be you.

Subjects were also asked the same question, but they were asked to respond on a *linear scale* (by circling one "+" point on the scale shown in Figure 4.1).

Subjects were asked the same question again, but they were asked to respond on a *log-scale* (by circling one "+" point on the scale shown in Figure 4.2

Finally, subjects were asked to respond to the same question and to answer on a log scale, but for *all of the years separately* in the 1980s (see Figure 4.3).

4.7.1 Consistency of Response

Subject's responses on the three different scales were compared for consistency. When the subject was found to be inconsistent on at least one scale, that subject was eliminated from the panel. This was done to avoid confounding the effects of inadequate quantitative training with substantive degree of belief. One hundred and seven subjects remained; these subjects were experts who responded consistently on all three scales. In Figure 4.4 we see the mean assessed probability as a function of year in the next decade. We also see the mean 90% credibility bands (assessed).

A sure thing ⇒	100%	+	1.00	⇒ Certain
		+		
	95%	+	.95	
		+		
	90%	+	.90	⇒ Very likely
		+		
	85%	+	.85	—
		+		
	80%	+	.80	
		+		
	75%	+	.75	
		+		⇒ Somewhat likely
	70%	+	.70	
		+		
	65%	+	.65	
		+		
	60%	+	.60	—
		+		
	55%	+	.55	
		+		
	50%	+	.50	⇒ Chances about even
		+		
	45%	+	.45	
		+		
	40%	+	.40	—
		+		
	35%	+	.35	
		+		
	30%	+	.30	
		+		⇒ Somewhat unlikely
	25%	+	.25	
		+		
	20%	+	.20	
		+		—
	15%	+	.15	
		+		
	10%	+	.10	⇒ Very unlikely
		+		
	5%	+	.05	
		+		
	0%	+	0.00	⇒ Impossible

Figure 4.1. Linear scale.

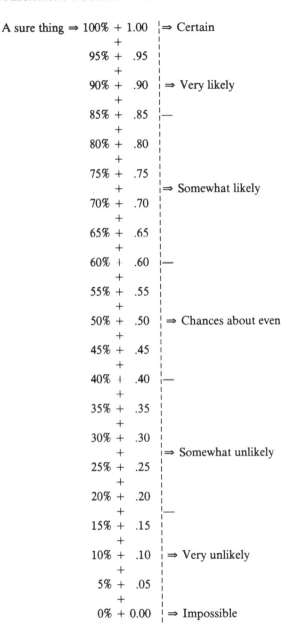

A sure thing →	1 +		The probability of:
	+		
	1/3 +		
	+		
A 10% chance →	1/10 +	←	a randomly chosen American
	+		being black.
	1/30 +		
	+		
A 1% chance →	1/100 +		
	+		
	1/300 +	←	a randomly chosen American
	+		male age 35–44 dying
One chance in a thousand →	1/1,000 +		during a 1-year period.
	+		
	1/3,000 +		
	+	←	a randomly chosen American
	1/10,000 +		being killed in a motor
	+		vehicle accident during a
	1/30,000 +		1-year period.
	+		
	1/100,000 +	←	getting all "heads" while
	+		flipping a fair coin 17
	1/300,000 +		times.
	+		
One chance in a million →	1/1,000,000 +		
	+	←	being killed in an
	1/3,000,000 +		accident during a single
	+		scheduled domestic airline
	1/10,000,000 +		flight.
	+	←	selecting a given
	1/30,000,000 +		Californian from among all
	+		Californians.
	1/100,000,000 +		
	+	←	a randomly chosen American
	1/300,000,000 +		will be you.
	+		
One chance in a billion →	1/1,000,000,000 +		
Less than	1/1,000,000,000 +		

Figure 4.2. Log scale.

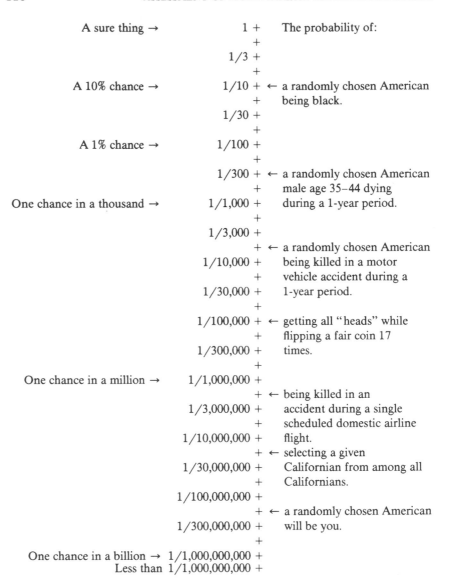

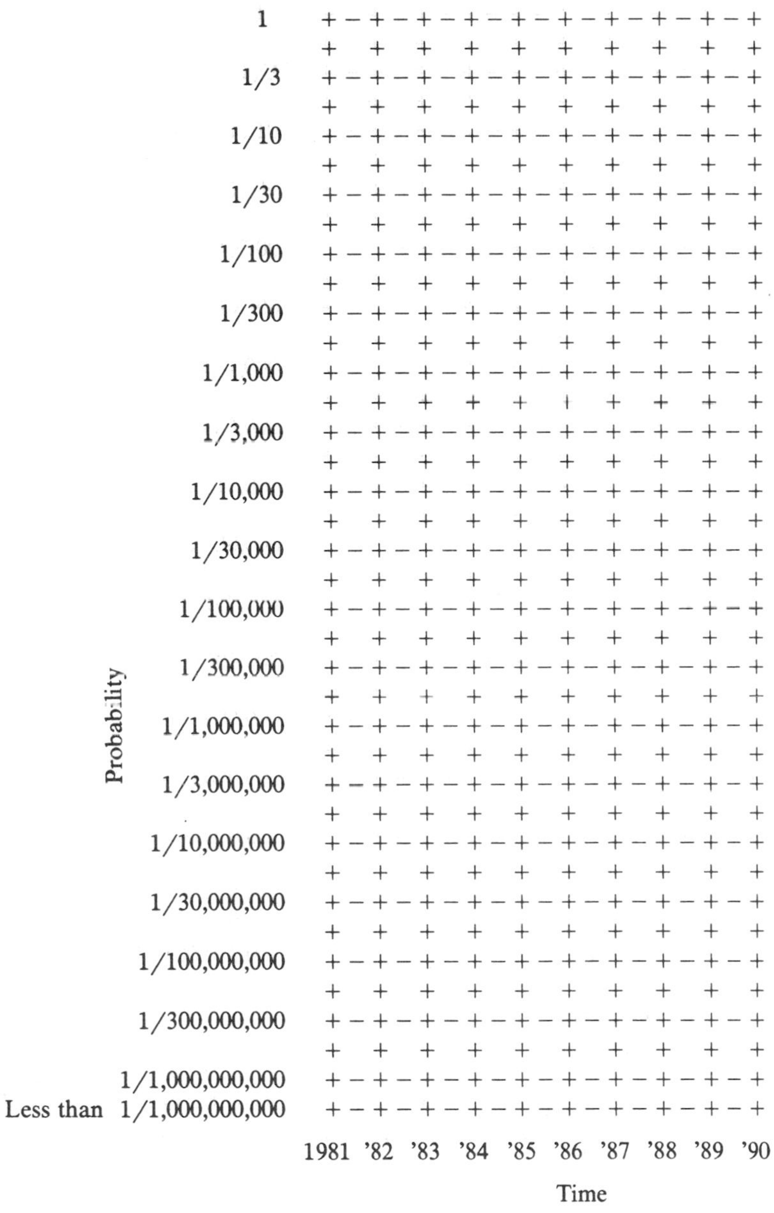

Figure 4.3. Log scale for all years separately.

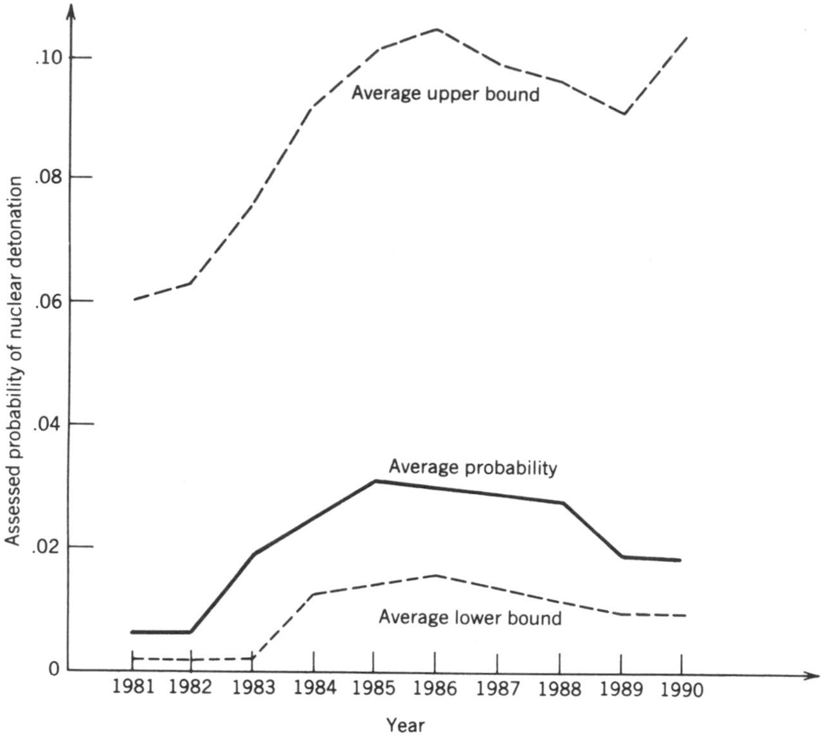

Figure 4.4. Assessments of consistent experts ($N = 107$).

4.7.2 Implications

1. The average (mean) assessed probability never exceeds about 3%.
2. The median assessed probability, in sharp contrast, never exceeds about .01%.
3. If the estimates of the probabilities for each year were independent (which they are not), a probability of .03 (3%) per year, for each year in the decade, would imply, using the binomial theorem, that the probability of war in the decade (i.e., the probability of exactly one "success" in 10 trials) is about 22.8%.

4.7.3 Histograms

In Figures 4.5–4.7 we see histograms for the years 1984, 1987, and 1990.

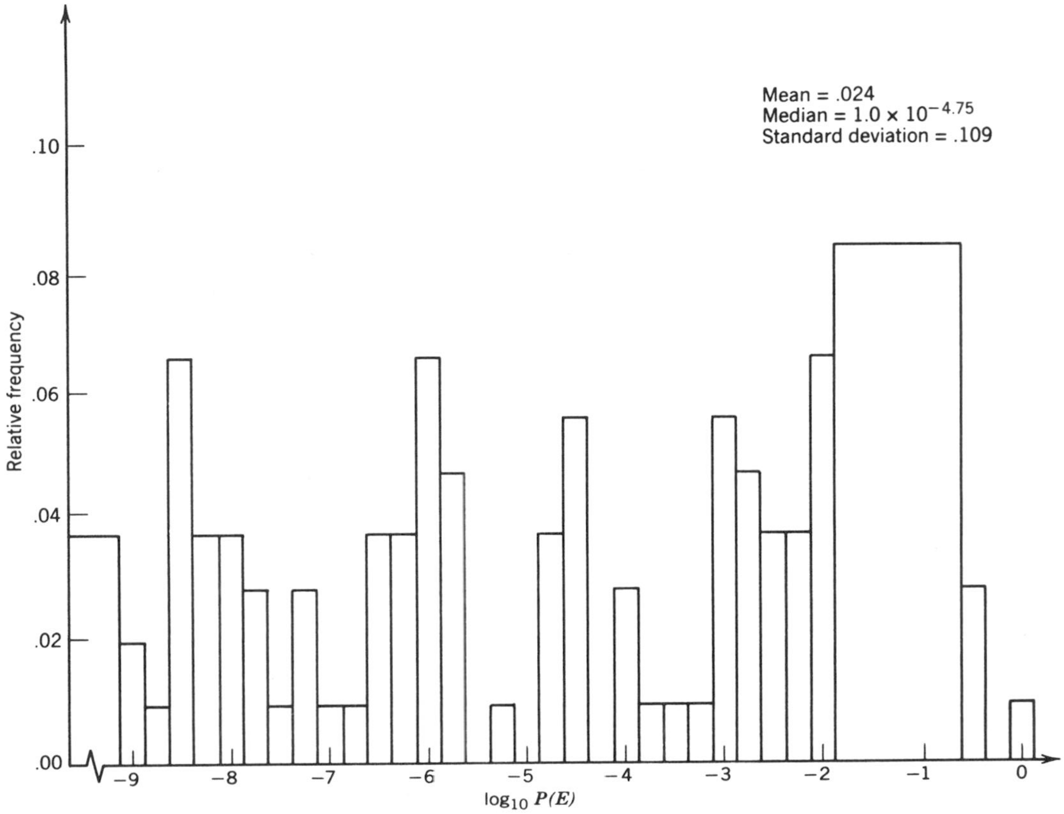

Figure 4.5. Histogram for assessed probability of nuclear detonation for 1984 ($N = 107$).

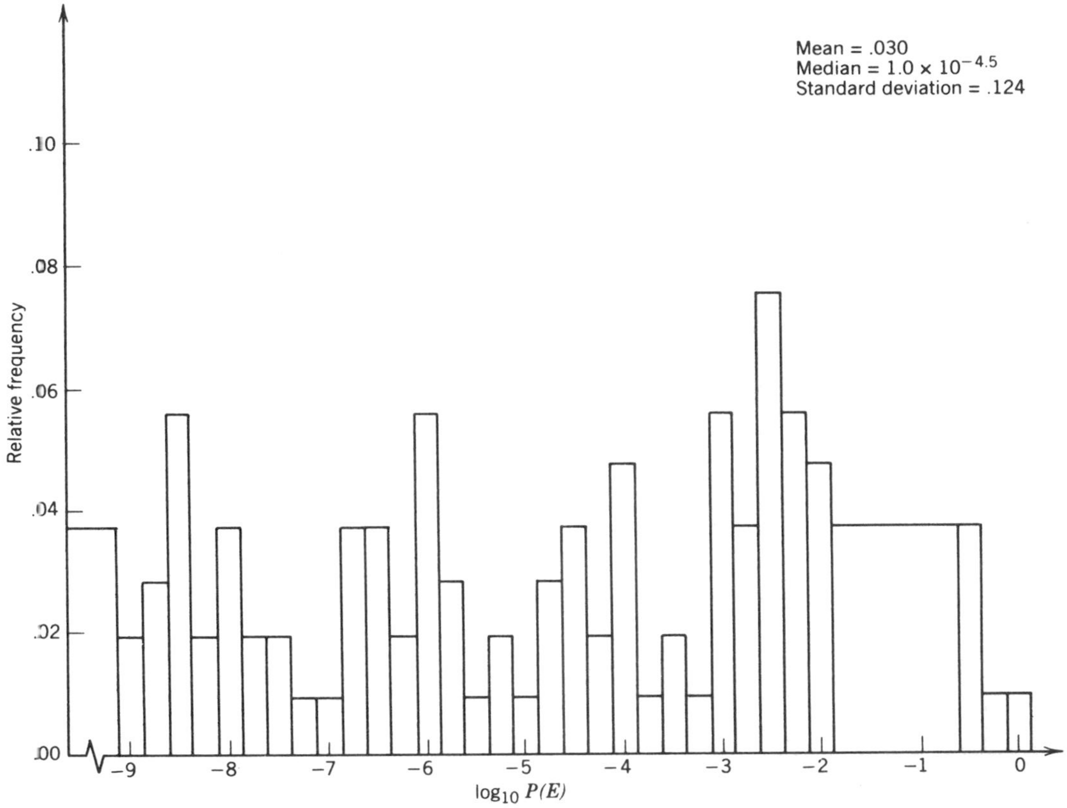

Figure 4.6. Histogram for assessed probability of nuclear detonation, 1987 ($N = 107$).

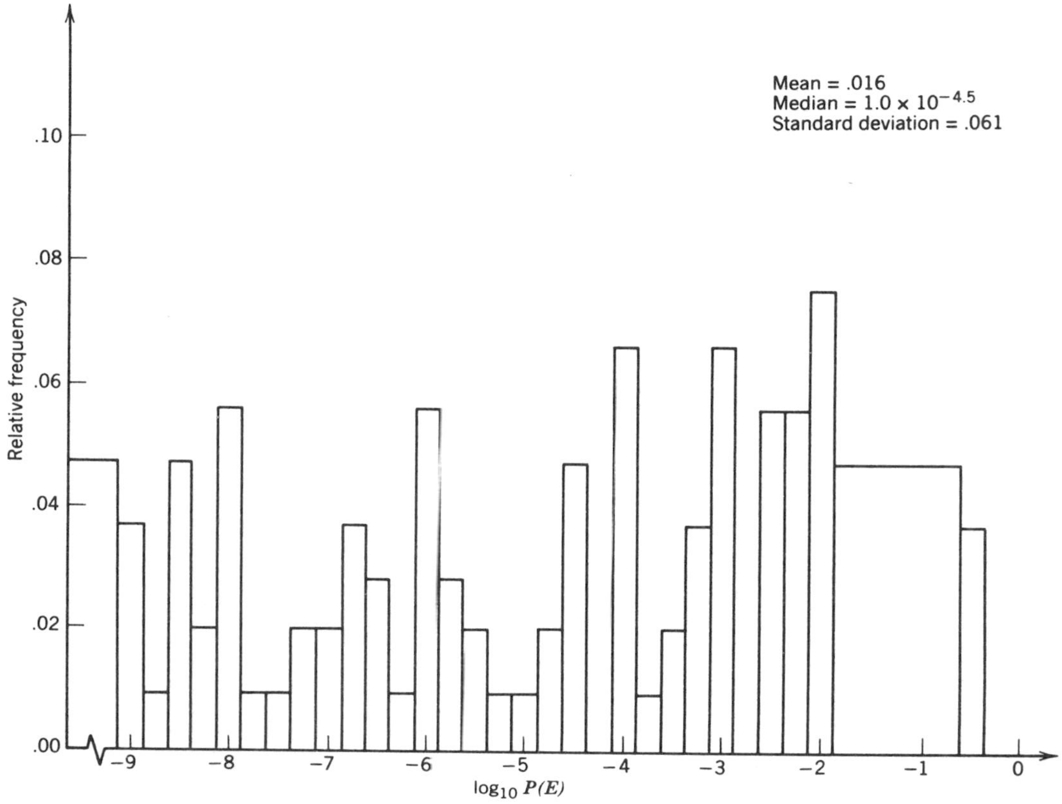

Figure 4.7. Histogram for assessed probability of nuclear detonation, 1990 (N = 107).

4.7.4 Smoothed Prior Density (Fitted)

Define $\boldsymbol{\theta}_j$: (10×1) as the response vector for panelist j; each component of $\boldsymbol{\theta}$ is the assessed probability for each of the 10 years in the decade, from 1981 through 1990. We assume $\boldsymbol{\theta}_j$ is an observation of $\boldsymbol{\theta}$, a continuous random vector with density $g(\boldsymbol{\theta})$; $j = 1, \ldots, 107$. The empirical density of responses (the prior density) is given by

$$g_{107}(\boldsymbol{\theta}) = \frac{1}{C^{10}(107)^{1/11}} \sum_{j=1}^{107} \frac{\exp\left\{\left[-(107)^{2/11}/2C^2\right](\boldsymbol{\theta} - \boldsymbol{\theta}_j)'(\boldsymbol{\theta} - \boldsymbol{\theta}_j)\right\}}{(2\pi)^5}.$$

This is a 10-dimensional density. The one-dimensional marginal density for

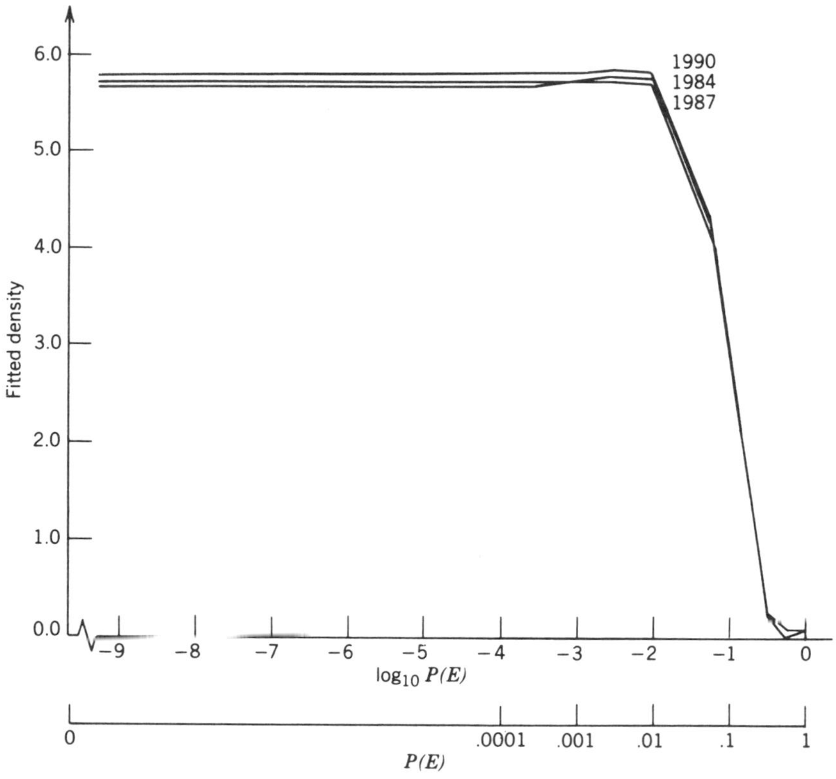

Figure 4.8. Fitted density for assessed probability of nuclear detonation, 1984, 1987, 1990 ($N = 107$) (log scale).

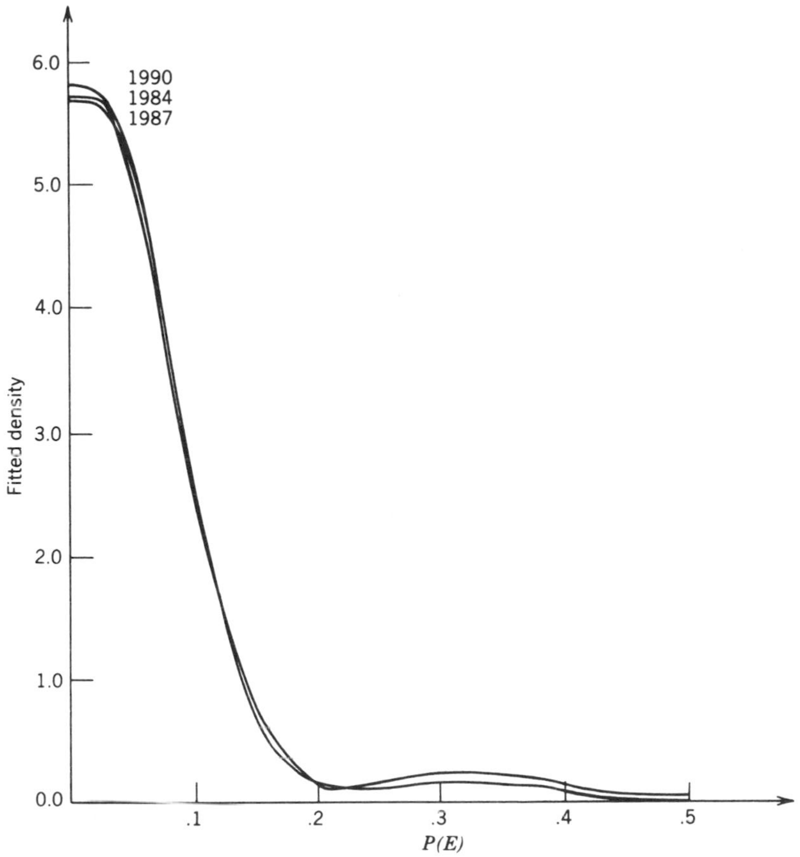

Figure 4.9. Fitted density for assessed probability of nuclear detonation, 1984, 1987, 1990 (N = 107) (linear scale).

the ith component is given by

$$g_{107}(\theta^{(i)}) = \frac{1}{C(107)^{10/11}} \sum_{j=1}^{107} \frac{\exp\left\{\left[-(107)^{2/11}/2C^2\right]\left(\theta^{(i)} - \theta_j^{(i)}\right)^2\right\}}{\sqrt{2\pi}}.$$

Plots of this density are given in Figures 4.8–4.11. Table 4.1 lists the three most frequently given reasons for the event of nuclear war.

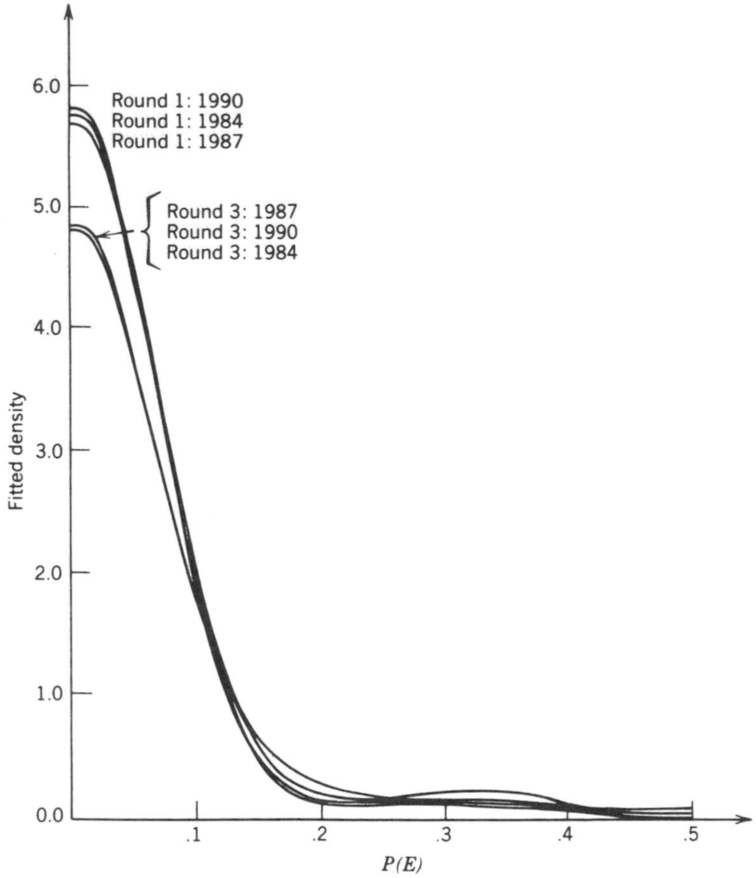

Figure 4.10. Fitted density for assessed probability of nuclear detonation. Round 1 ($N = 107$): consistent experts. Round 3 ($N = 82$).

Note that the smoothing constant $C = 0.1$. This choice yielded a smooth, well behaved density on a scale that was readily interpretable (the density became extremely noisy for smaller C, and it was overfitted for larger C). The assessment was carried out with both experts (Figures 4.9 and 4.10) and non-experts (Figure 4.11). Figs. 4.10 and 4.11 compare the density of responses of a group of 107 consistent experts (and non-experts) with the third round responses of the 82 consistent experts who remained in the study after we repeatedly asked the group the same questions three times and fed back to them after each questioning a composite of reasons given by the group.

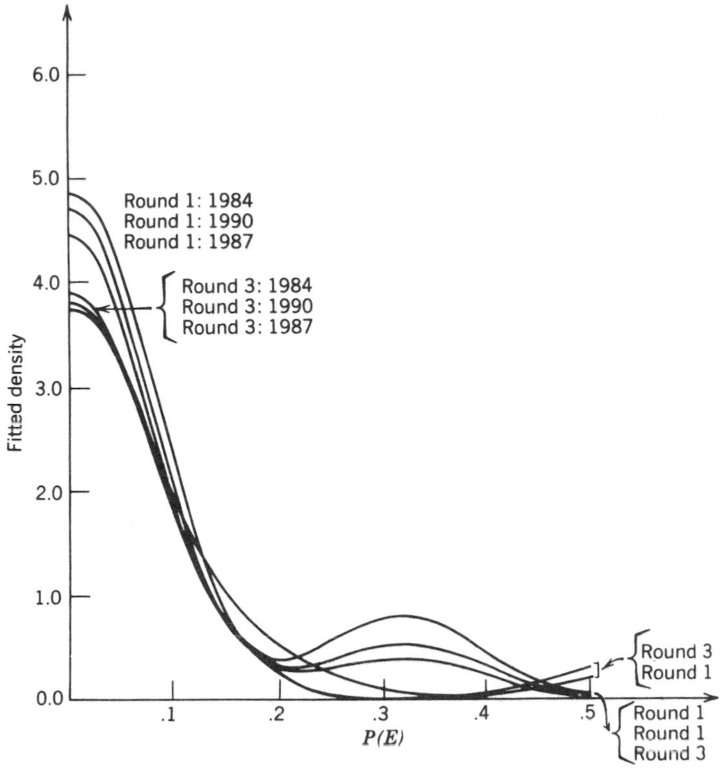

Figure 4.11. Fitted density for assessed probability of nuclear detonation. Round 1 ($N = 20$): consistent nonexperts. Round 3 ($N = 13$).

4.8 PSYCHOLOGICAL FACTORS RELATING TO SUBJECTIVE PROBABILITY ASSESSMENT

There are many biases in the subjective assessments made by individuals (for additional discussion of this topic see Tversky and Kahneman, 1974; Hogarth, 1980; and Kahneman, Slovic, and Tversky, 1982). A few major biases are listed below.

BIASES

1. *Desirable / Undesirable Events.* Subjects tend to underassess the probability of undesirable events (nuclear war).
2. *Availability Heuristic.* Subjects tend to underassess the probability of events for which they cannot quickly construct previous similar

Table 4.1 Most Frequently Given Reasons for the Event of Nuclear War

Reason	Response Frequency	Response Frequency rank
1. Increasing world nuclear proliferation increases the chances of nuclear war among other third-party nations facing critical economic, political, and demographic problems, and it makes third parties much more likely to get into direct nuclear conflict than the United States and the Soviet Union	14 %	1
2. The United States and the Soviet Union both view Middle East oil as vital to their economies and will find their oil contention peaking over the next decade, causing them to become more directly involved in unstable Third-World countries, perhaps using military force to secure supplies.	13 %	2
3. Probability of accident (computer, radar, or human failure) causing a nuclear strike is greater than the probability of the United States or the Soviet Union intentionally striking the other side.	12.9%	3

occurrences or scenarios, and certainly for events that have never occurred.

3. *Low / High Probability Events.* Subjects tend to overassess extremely low probability events because they have only incomplete understanding of how to distinguish among such infrequently occurring events and do not have real appreciation of their rarety (such as the rare event of nuclear war).
4. *Anchoring Heuristic.* Subjects tend to underassess disjunctive events, such as events that cause failure of a complex system that tends to break down when any of its really essential components fail (such as systems of safeguards against nuclear war). Moreover, they are unlikely to substantially change an intellectual position they have espoused.

Conclusions Regarding Psychological Factors

Some of the psychologically related sources of potential bias in the subjective assessments lead to overassessment, but most appear to lead to underassessment. There has been very little research on how to evaluate the effects of competing heuristics. We conclude that the above assessments are probably somewhat too low, and we view the results as conservative. Much more research remains to be carried out in this area.

4.9 SUMMARY

In this chapter we have focused on the problem of how to assess multivariate prior distributions. We have suggested that in many contexts it should be reasonable to substitute (a) the distribution of the p-point assessments of a random vector for N individuals for (b) the p-dimensional distribution of an unknown vector for a single individual. In such cases a normally very hard problem is very easily solved. We have applied this procedure to the problem of finding a prior distribution for the probability of nuclear war between the United States and the Soviet Union over the decade of the 1980s.

EXERCISES

4.1* In assessing a multivariate prior distribution by assessing a p-dimensional vector from N subjects and merging them by density estimation, we used a kernel, $K(\boldsymbol{\theta})$, which we took to be normal. Suppose instead that we choose $K(\boldsymbol{\theta})$ to be multivariate Student's t, so that

$$K(\boldsymbol{\theta}) \propto \{n + \boldsymbol{\theta}'\boldsymbol{\theta}\}^{-(n+p)/2},$$

for some suitable n. How would this change be likely to affect the usefulness of the fitted density of opinions? (*Hint:* See literature on density estimation such as Taipia and Thompson, 1982.)

4.2 How should an expert be defined?

4.3 Suppose you are trying to decide whether to buy shares in the XYZ corporation. You seek an opinion from several of the well-known stock brokerage firms in the closest major financial center. One decision rule is to follow the advice of the majority of opinions. Another might involve finding out how confident each firm is in its opinion. How would you generalize this concept to the multivariate prior distribution assessment context? (*Hint*: See Section 4.6, paragraph 4.)

*Asterisked exercises require reference to sources outside of this text. Full reference information can be found in the bibliography at the back of this book.

4.4 Compare the mean assessed probability with the median assessed probability in Section 4.7. How do you interpret the substantial discrepancy?

4.5 By using the availability heuristic (see Section 4.8), explain why many people do not perceive the advent of a nuclear war in the near future as a likely event.

4.6* Give another method that has been proposed for assessing multivariate prior distributions. (*Hint*: See Kadane et al., 1980.)

PART 2

Models and Applications

CHAPTER V

Bayesian Inference in Regression

5.1 INTRODUCTION

This chapter presents some basic applications of Bayesian analysis to problems of interest in regression. These problems include the univariate and multivariate linear regression models. We treat only the simplest of cases, namely, uncorrelated observational data with equal variances. This approach is taken to keep the exposition simple, so that the Bayesian methodology is not obfuscated by model complexity. More complex regression models such as models with correlated disturbances, models with heteroscedasticity, multiple linear regression, and time series models are treated from a Bayesian point of view elsewhere (see, e.g., Broemeling, 1985; Leamer, 1978; Press, 1982; and Zellner, 1971 and 1985). Missing observations in regression are studied from a Bayesian point of view by Rubin (1987). He develops there a Bayesian method for imputing the missing values; the method is called *multiple imputation*. Bayesian inference in factor analysis is discussed in Press and Shigemasu, 1989.

5.2 SIMPLE LINEAR REGRESSION

5.2.1 Model

The "simple" linear regression model relates a single dependent variable to a single independent variable, linearly in the coefficients. The model is called "simple" if there is only one independent variable. It is called "multiple" if there is more than one independent variable. The model is

$$y_i | x_i = \beta_1 + \beta_2 x_i + u_i, \qquad i = 1, \ldots, n,$$

where y_i denotes the ith observation on the dependent variable, x_i denotes the ith observation on the independent variable, u_i denotes the ith disturbance or error, and (β_1, β_2) are unknown parameters.

[*Note:* Assume $u_1, \ldots, u_n$ are independent $N(0, \sigma^2)$, so that the unknown parameters of the model are $(\beta_1, \beta_2, \sigma^2)$.]

Likelihood Function

Let $\mathbf{y} = (y_1, \ldots, y_n)'$, and let $\mathbf{x} = (x_1, \ldots, x_n)'$. Ignoring a proportionality constant, the likelihood function becomes

$$L(\mathbf{y}|\mathbf{x}, \beta_1, \beta_2, \sigma) = \frac{1}{\sigma^n} \exp\left\{ \frac{1}{2\sigma^2} \sum_1^n (y_i - \beta_1 - \beta_2 x_i)^2 \right\}.$$

Prior

Suppose we adopt the vague prior density

$$g(\beta_1, \beta_2, \sigma) = g_1(\beta_1) g_2(\beta_2) g_3(\sigma),$$

$$g_1(\beta_1) \propto \text{constant},$$

$$g_2(\beta_2) \propto \text{constant},$$

$$g_3(\sigma) \propto \frac{1}{\sigma}.$$

(Note that an alternative prior that could be used is the g-prior; see Section 2.7.3.) Then, by Bayes' theorem, the joint posterior density is

$$h(\beta_1, \beta_2, \sigma|\mathbf{x}, \mathbf{y}) \propto \frac{1}{\sigma} \cdot \frac{1}{\sigma^n} \exp\left\{ -\frac{1}{2\sigma^2} \sum_1^n (y_i - \beta_1 - \beta_2 x_i)^2 \right\},$$

$$-\infty < \beta_1, \beta_2 < +\infty, 0 < \sigma < \infty.$$

Equivalently,

$$h(\beta_1, \beta_2, \sigma|\mathbf{x}, \mathbf{y}) \propto \frac{1}{\sigma^{n+1}} \exp\left\{ -\frac{1}{2\sigma^2} \sum_1^n (y_i - \beta_1 - \beta_2 x_i)^2 \right\}.$$

The MLE's are

$$\hat{\beta}_1 = \bar{y} - \hat{\beta}_2 \bar{x}, \qquad \hat{\beta}_2 = \frac{\sum (x_i - \bar{x})(y_i - \bar{y})}{\sum (x_i - \bar{x})^2},$$

where: $\bar{x} = n^{-1}\Sigma x_i$, $\bar{y} = n^{-1}\Sigma y_i$, and an unbiased estimator of σ^2 is

$$\hat{\sigma}^2 = s^2 = \frac{1}{n-2}\sum\left(y_i - \hat{\beta}_1 - \hat{\beta}_2 x_i\right)^2.$$

Writing

$$\sum_1^n \left(y_i - \beta_1 - \beta_2 x_i\right)^2$$
$$= \sum_1^n \left[\left(y_i - \hat{\beta}_1 - \hat{\beta}_2 x_i\right) - \left(\beta_1 - \hat{\beta}_1\right) - \left(\beta_2 - \hat{\beta}_2\right)x_i\right]^2,$$

combined with integrating with respect to σ, gives the two-dimensional posterior density

$$h_1(\beta_1, \beta_2 | \mathbf{y}, \mathbf{x})$$
$$\propto \left\{(n-2)s^2 + n\left(\beta_1 - \hat{\beta}_1\right)^2 + \left(\beta_2 - \hat{\beta}_2\right)^2 \sum x_i^2 + 2\left(\beta_1 - \hat{\beta}_1\right)\left(\beta_2 - \hat{\beta}_2\right)\sum x_i\right\}^{-n/2}.$$

That is, β_1 and β_2, given $\mathbf{x}$ and $\mathbf{y}$, jointly follow a bivariate Student's t-distribution, so that marginally they each follow univariate Student's t-posterior distributions (see, e.g., Press, 1982, Section 6.2).

Marginally, we have

$$\left[\frac{\sum (x_i - \bar{x})^2}{\frac{s^2}{n}\sum x_i^2}\right]^{1/2} \cdot \left(\beta_1 - \hat{\beta}_1\right) | \mathbf{y}, \mathbf{x} \sim t_{n-2}$$

and

$$\left.\frac{\beta_2 - \hat{\beta}_2}{\left(\frac{s}{\left[\sum (x_i - \bar{x})^2\right]^{1/2}}\right)}\right| \mathbf{y}, \mathbf{x} \sim t_{n-2}.$$

For example, we can make credibility statements about the slope coefficient β_2 (see Figure 5.1):

$$P\left\{\hat{\beta}_2 - z_\alpha \delta \le \beta_2 \le \hat{\beta}_2 + z_\alpha \delta | \mathbf{y}, \mathbf{x}\right\} = 1 - 2\alpha,$$

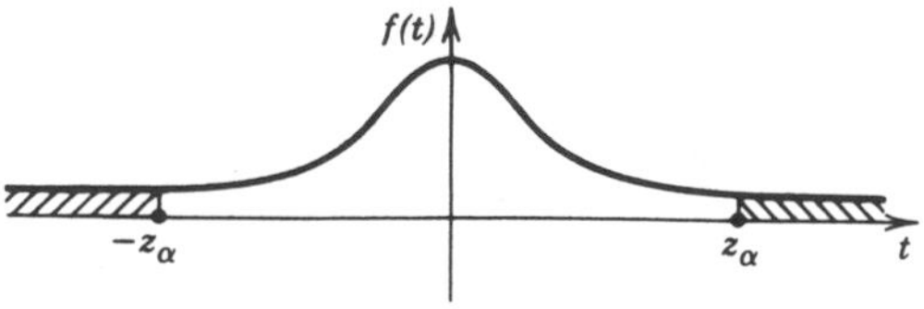

Figure 5.1. Posterior density of β_2.

where z_α is the α-fractile point of the Student's t-distribution, $f(t)$ is its density, and

$$\delta \equiv \frac{s}{\left[\sum (x_i - \bar{x})^2\right]^{1/2}}.$$

For example, suppose y_i denotes crop yield in year i, and x_i denotes quantity of fertilizer used in year i, and we have data for 20 years ($n = 20$) that yield the estimators (in appropriate units)

$$\hat{\beta}_1 = 345, \qquad \hat{\beta}_2 = 3.00, \qquad s^2 = 662.8,$$

$$\bar{x} = 45.35, \qquad \frac{1}{n}\sum (x_i - \bar{x})^2 = 285.55.$$

(These data are adapted from an example in Zellner (1971), p. 64, that was in turn developed from Haavelmo, 1947.) Then, the (marginal) posterior densities are given in Figures 5.2 and 5.3. The posterior means are equal to the MLE's and are given by

$$E(\beta_1 | \mathbf{y}, \mathbf{x}) = 345, \qquad E(\beta_2 | \mathbf{y}, \mathbf{x}) = 3.00.$$

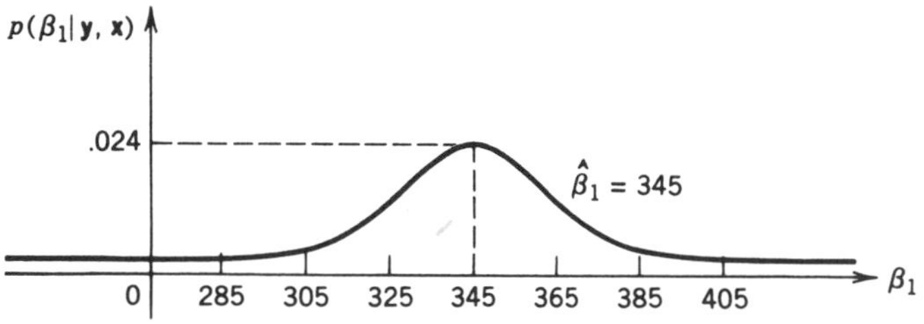

Figure 5.2. Posterior density of β_1.

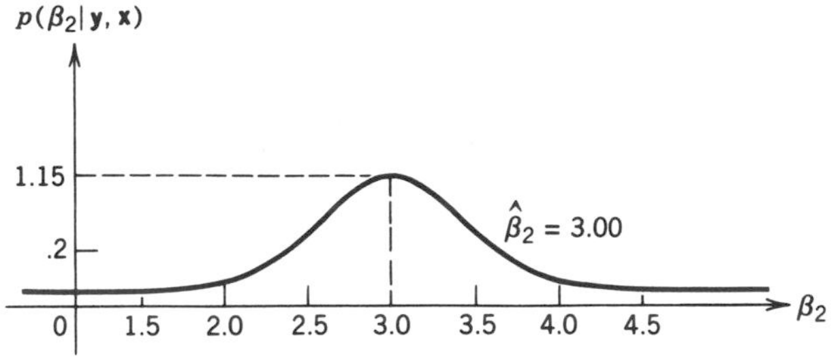

Figure 5.3. Posterior density of β_2.

Moreover,

$$P\{1.5 \leq \beta_2 \leq 4.5|\mathbf{y}, \mathbf{x}\} \cong 1 \qquad (P > 99.95\%).$$

We use these types of results to make predictions of yields.

5.2.2 Predictive Distribution

Suppose we wish to find predictive intervals for a new output y^*, based upon a new observation x^* and past experience. We know

$$y^*|x^* = \beta_1 + \beta_2 x^* + u, \qquad u \sim N(0, \sigma^2).$$

The predictive density of y^* is given by

$$p(y^*|x^*) = \int\int\int L(y^*|x^*, \beta_1, \beta_2, \sigma) h(\beta_1, \beta_2, \sigma|\mathbf{x}, \mathbf{y})\, d\beta_1, d\beta_2, d\sigma.$$

It is straightforward to show [see, e.g., Press (1982, p. 258)] that

$$\left.\frac{y^* - \hat{\beta}_1 - \hat{\beta}_2 x^*}{\hat{\sigma}\left[1 + \dfrac{1}{n} + \dfrac{(x^* - \bar{x})^2}{\Sigma_1^n (x_i - \bar{x})^2}\right]^{1/2}}\right| \text{sample} \sim t_{n-2}.$$

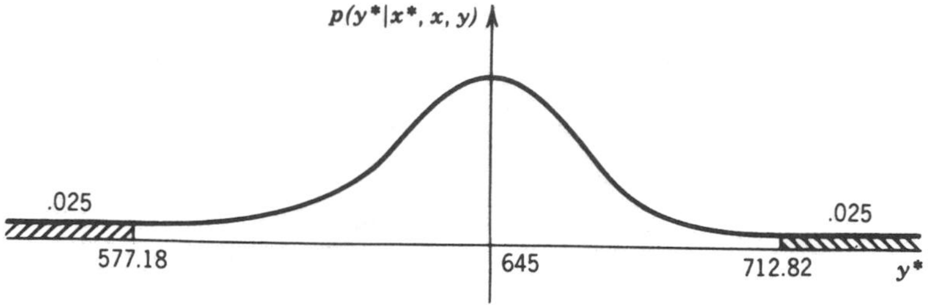

Figure 5.4. Predictive density.

Assuming $x^* = 100$, the data from the example in Section 5.2.1 gives

$$\left.\frac{y^* - 645}{32.28}\right| \text{sample} \sim t_{18}.$$

At a credibility level of 95%, this gives (see Figure 5.4)

$$P\{577.18 \leq y^* \leq 712.82 | \text{sample}\} = 95\%.$$

We can, of course, find smaller intervals by lowering the credibility level to less than 95%.

5.2.3 Posterior Inferences About the Standard Deviation

It is straightforward to make posterior inferences about σ from the marginal posterior density. It is given by (using h generically)

$$h(\sigma|\mathbf{x},\mathbf{y}) = \int\int h(\beta_1, \beta_2, \sigma|\mathbf{x}, \mathbf{y})\, d\beta_1\, d\beta_2;$$

or, using the functional form given in Section 5.2.1, we find for the marginal posterior density of σ,

$$h(\sigma|\mathbf{x}, \mathbf{y}) \propto \int\int \frac{1}{\sigma^{n+1}} e^{-(1/2\sigma^2)q(\beta_1, \beta_2)}\, d\beta_1\, d\beta_2,$$

where

$$q(\beta_1, \beta_2) \equiv (n-2)s^2 + n(\beta_1 - \hat{\beta}_2)^2 + (\beta_2 - \hat{\beta}_2)\Sigma x_i^2$$
$$+2(\beta_1 - \hat{\beta}_1)(\beta_2 - \hat{\beta}_2)\Sigma x_i.$$

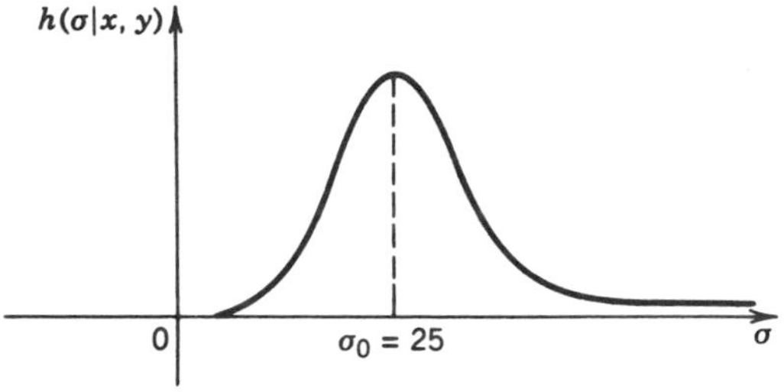

Figure 5.5. Posterior density of σ.

Carrying out the integration gives (see Figure 5.5)

$$h(\sigma|\mathbf{x},\mathbf{y}) \propto \frac{1}{\sigma^{n-1}} e^{-(n-2)s^2/2\sigma^2}, \qquad 0 < \sigma < \infty, 2 < n,$$

an inverted gamma distribution density (i.e., $1/\sigma$ follows a gamma distribution). Thus,

$$E(\sigma|\mathbf{x},\mathbf{y}) = \frac{\Gamma\left(\frac{n-3}{2}\right)}{\Gamma\left(\frac{n-2}{2}\right)} \left(\frac{n-2}{2}\right)^{1/2} s,$$

$$\operatorname{var}(\sigma|\mathbf{x},\mathbf{y}) = \frac{(n-2)s^2}{(n-4)} - [E(\sigma|\mathbf{x},\mathbf{y})]^2.$$

Furthermore,

$$\sigma_0 = \text{modal value of } \sigma = s\sqrt{\frac{n-2}{n-1}}$$

for

$$s^2 = 662.8, \qquad n = 20, \qquad \sigma_0 = 25.$$

5.3 MULTIVARIATE REGRESSION MODEL

Let $\mathbf{V}$ denote a $p \times p$, symmetric, positive definite matrix of random variables, and we write $\mathbf{V} > 0$. This means that $\mathbf{V} = \mathbf{V}'$ and that all of the latent roots of **V** are positive. **V** follows a Wishart distribution (see also Section 3.4) if the elements are jointly continuous with multivariate probability density function:

$$p(\mathbf{V}|\Sigma, n) \propto \frac{|\mathbf{V}|^{(n-p-1)/2}}{|\Sigma|^{n/2}} e^{-\operatorname{tr} \Sigma^{-1}\mathbf{V}/2}, \qquad p \le n,\ \Sigma > 0,$$

and we write

$$\mathbf{V} \sim W(\Sigma, p, n).$$

Note that Σ is a scale matrix, n is a degrees-of-freedom parameter, and p is the dimension. The proportionality constant depends upon (p, n) only.

5.3.1 Multivariate Vague Priors

Suppose we want to express indifference about the values of the elements of Σ with a vague prior. We would use the improper prior density

$$p(\Sigma) \propto \frac{1}{|\Sigma|^{(p+1)/2}}.$$

This prior is obtainable (see Geisser and Cornfield, 1963) from the structural form proposed by Jeffreys (1961), who used an argument based upon invariance. It is also the prior implied by fiducial probability (Villegas, 1969) and by structural probability (Fraser, 1968). For $p = 1$ it reduces to the conventional univariate vague prior. For additional discussion of multivariate priors see, for example, Press (1982, 1984).

5.3.2 Multivariate Regression

The multivariate regression model relates p correlated dependent variables to q independent variables using linear relationships. The model is

$$\underset{(N\times 1)}{\mathbf{y}_1} |\mathbf{X} = \underset{(N\times q)}{\mathbf{X}} \cdot \underset{(q\times 1)}{\boldsymbol{\beta}_1} + \underset{(N\times 1)}{\mathbf{u}_1},$$

$$\vdots \qquad\qquad \vdots \qquad\qquad \vdots$$

$$\underset{(N\times 1)}{\mathbf{y}_p} |\mathbf{X} = \underset{(N\times q)}{\mathbf{X}} \cdot \underset{(q\times 1)}{\boldsymbol{\beta}_p} + \underset{(N\times 1)}{\mathbf{u}_p}$$

where

$$\mathbf{u}_j \sim N(\mathbf{0}, \sigma_{jj}\mathbf{I}_N), \qquad j = 1, \ldots, p,$$

and the $\mathbf{u}_j$'s are correlated. Evidently, if

$$\mathbf{Y} \equiv (\mathbf{y}_1, \ldots, \mathbf{y}_p), \qquad \mathbf{U} \equiv (\mathbf{u}_1, \ldots, \mathbf{u}_p), \qquad \mathbf{B} = (\boldsymbol{\beta}_1, \ldots, \boldsymbol{\beta}_p),$$

the model becomes

$$\underset{(N\times p)}{\mathbf{Y}} = \underset{(N\times q)}{\mathbf{X}} \underset{(q\times p)}{\mathbf{B}} + \underset{(N\times p)}{\mathbf{U}},$$

$$\mathbf{U}' \equiv (\mathbf{v}_1, \ldots, \mathbf{v}_N),$$

$$\mathbf{v}_j \sim N(\mathbf{0}, \boldsymbol{\Sigma}), \qquad \text{i.i.d., } \boldsymbol{\Sigma} > 0.$$

Thus, a multivariate regression model involves the study of several univariate regressions taken simultaneously, because the disturbance terms in the several regressions are mutually correlated. So information in one regression can be used to estimate coefficients in other regressions (and the same is true for predictions of future response variable values). Bayesian inference in the multivariate regression model involves use of the multivariate normal distribution, as well as the Wishart distribution, just defined.

Likelihood Function

The likelihood function of $\mathbf{U}'$ (same as that of $\mathbf{U}$) is the joint density of the $\mathbf{v}_j$'s and, by independence, is given by

$$p(\mathbf{U}|\boldsymbol{\Sigma}) = p(\mathbf{v}_1, \ldots, \mathbf{v}_N|\boldsymbol{\Sigma}) = \prod_1^N p(\mathbf{v}_j|\boldsymbol{\Sigma}),$$

or

$$p(\mathbf{U}|\boldsymbol{\Sigma}) \propto \frac{1}{|\boldsymbol{\Sigma}|^{N/2}} e^{-\operatorname{tr} \boldsymbol{\Sigma}^{-1}\mathbf{U}'\mathbf{U}/2},$$

since $\mathbf{U}'\mathbf{U} = \Sigma_1^N \mathbf{v}_j\mathbf{v}_j'$. Since $(\mathbf{Y}|\mathbf{X}, \mathbf{B}) = \mathbf{XB} + \mathbf{U}$, and since the Jacobian of the transformation* is unity, transforming variables from $\mathbf{U}$ to $\mathbf{Y}$ gives

$$p(\mathbf{Y}|\mathbf{X}, \mathbf{B}, \boldsymbol{\Sigma}) \propto \frac{1}{|\boldsymbol{\Sigma}|^{N/2}} e^{-\operatorname{tr} \boldsymbol{\Sigma}^{-1}(\mathbf{Y}-\mathbf{XB})'(\mathbf{Y}-\mathbf{XB})/2}.$$

*The Jacobian of the one-to-one transformation $\mathbf{Y} = f(\mathbf{X})$, where the distinct elements of $\mathbf{Y}$ are $(y_1, \ldots, y_n)$, and where those of $\mathbf{X}$ are $(x_1, \ldots, x_n)$, is $|\det(\mathbf{A})|^{-1}$, where $\mathbf{A} \equiv (a_{ij})$ and $a_{ij} = \partial y_i / \partial x_j$; $\det(\mathbf{A})$ denotes determinant of the matrix $\mathbf{A}$. In this case, $\mathbf{Y}$ and $\mathbf{U}$ are related by an additive constant, so the Jacobian is unity.

Orthogonality Property of Least-Squares Estimators. Let $\hat{\mathbf{B}}$ denote the least squares estimator. By adding and subtracting $\mathbf{X}\hat{\mathbf{B}}$, we can find

$$\begin{aligned}\mathbf{A} &\equiv (\mathbf{Y} - \mathbf{X}\mathbf{B})'(\mathbf{Y} - \mathbf{X}\mathbf{B}) \\ &= \left[(\mathbf{Y} - \mathbf{X}\hat{\mathbf{B}}) + \mathbf{X}(\hat{\mathbf{B}} - \mathbf{B})\right]'\left[(\mathbf{Y} - \mathbf{X}\hat{\mathbf{B}}) + \mathbf{X}(\hat{\mathbf{B}} - \mathbf{B})\right].\end{aligned}$$

But it is easy to check that the least-squares estimator is

$$\hat{\mathbf{B}} = (\mathbf{X}'\mathbf{X})^{-1}\mathbf{X}'\mathbf{Y}.$$

It may be found that $\mathbf{X}'\hat{\mathbf{U}} \equiv \mathbf{X}'(\mathbf{Y} - \mathbf{X}\hat{\mathbf{B}}) = 0$, the orthogonality property of least-squares estimators, so that the cross-product terms in ($\mathbf{A}$) vanish, and we get

$$\mathbf{A} = \mathbf{V} + (\mathbf{B} - \hat{\mathbf{B}})'(\mathbf{X}'\mathbf{X})(\mathbf{B} - \hat{\mathbf{B}}),$$

where:

$$\mathbf{V} \equiv (\mathbf{Y} - \mathbf{X}\hat{\mathbf{B}})'(\mathbf{Y} - \mathbf{X}\hat{\mathbf{B}}).$$

Then, the likelihood function becomes

$$p(\mathbf{Y}|\mathbf{X}, \mathbf{B}, \boldsymbol{\Sigma}) \propto \frac{1}{|\boldsymbol{\Sigma}|^{N/2}} e^{-\operatorname{tr} \boldsymbol{\Sigma}^{-1}[\mathbf{V} + (\mathbf{B} - \hat{\mathbf{B}})'(\mathbf{X}'\mathbf{X})(\mathbf{B} - \hat{\mathbf{B}})]/2}.$$

Vague Priors

Adopting a vague prior distribution on ($\mathbf{B}, \boldsymbol{\Sigma}$) gives

$$p(\mathbf{B}, \boldsymbol{\Sigma}) = p(\mathbf{B})p(\boldsymbol{\Sigma}),$$

$$p(\mathbf{B}) \propto \text{Constant},$$

$$p(\boldsymbol{\Sigma}) \propto \frac{1}{|\boldsymbol{\Sigma}|^{(p+1)/2}},$$

$$p(\mathbf{B}, \boldsymbol{\Sigma}) \propto \frac{1}{|\boldsymbol{\Sigma}|^{(p+1)/2}}.$$

Posterior Analysis

By Bayes' theorem, the joint posterior density is

$$p(\mathbf{B}, \boldsymbol{\Sigma}|\mathbf{X}, \mathbf{Y}) \propto \frac{1}{|\boldsymbol{\Sigma}|^{(N+p+1)/2}} e^{-\operatorname{tr} \boldsymbol{\Sigma}^{-1}\mathbf{A}/2}.$$

Posterior inferences about **B** are made most easily from the marginal posterior density, given by

$$p(\mathbf{B}|\mathbf{X},\mathbf{Y}) \propto \int_{\boldsymbol{\Sigma}>0} \frac{1}{|\boldsymbol{\Sigma}|^{(N+p+1)/2}} e^{-\operatorname{tr}\boldsymbol{\Sigma}^{-1}\mathbf{A}/2}\, d\boldsymbol{\Sigma}.$$

This integration may be carried out readily by recognizing that if $\boldsymbol{\Sigma}^{-1}$ followed a Wishart distribution, its density would be proportional to that in the integrand. The result is

$$p(\mathbf{B}|\mathbf{X},\mathbf{Y}) \propto \frac{1}{|\mathbf{V} + (\mathbf{B}-\hat{\mathbf{B}})'(\mathbf{X}'\mathbf{X})(\mathbf{B}-\hat{\mathbf{B}})|^{N/2}}.$$

That is, **B** follows a matrix **T**-distribution. Therefore its columns (and rows) marginally follow multivariate Student's **t**-distributions, so that for $j = 1,\ldots,p$ we have for the columns of **B**,

$$p(\boldsymbol{\beta}_j|\mathbf{X},\mathbf{Y}) \propto \frac{1}{\left\{v_{jj} + \left(\boldsymbol{\beta}_j - \hat{\boldsymbol{\beta}}_j\right)'(\mathbf{X}'\mathbf{X})\left(\boldsymbol{\beta}_j - \hat{\boldsymbol{\beta}}_j\right)\right\}^{(N-p+1)/2}},$$

where

$$\mathbf{V} \equiv (v_{ij}).$$

Joint posterior inferences about all of the coefficients in a given regression equation may be made from the result:

$$\left[\left(\boldsymbol{\beta}_j - \hat{\boldsymbol{\beta}}_j\right)'\mathbf{G}_j\left(\boldsymbol{\beta}_j - \hat{\boldsymbol{\beta}}_j\right)|\mathbf{X},\mathbf{Y}\right] \sim F_{q+1,\,N-p-q},$$

where

$$\mathbf{G}_j \equiv \frac{\mathbf{X}'\mathbf{X}(N-p-q)}{(N-p-q+1)(q+1)s_j^2}, \qquad s_j^2 = \frac{v_{jj}}{N-p-q+1}.$$

Posterior inferences about specific regression coefficients are made from the marginal univariate Student's *t*-distributions:

$$\left[\frac{\beta_{ij} - \hat{\beta}_{ij}}{s_j\sqrt{k_{ii}}}\middle|\hat{\beta}_{ij}, s_j, \mathbf{X}\right] \sim t_{N-p-q+1},$$

where β_{ij} denotes the ith element of $\boldsymbol{\beta}_j$, and

$$\mathbf{K} \equiv (\mathbf{X}'\mathbf{X})^{-1} = (k_{ij}).$$

5.3.3 Posterior Inferences About the Covariance Matrix

It is straightforward to check, by integrating the joint posterior density of $(\mathbf{B}, \boldsymbol{\Sigma} | \mathbf{X}, \mathbf{Y})$ with respect to $\mathbf{B}$ and by transforming $\boldsymbol{\Sigma} \to \boldsymbol{\Sigma}^{-1}$ (the Jacobian is $|\boldsymbol{\Sigma}|^{-(p+1)}$; see, for example, Press, 1982, p. 47), that the marginal density of $\boldsymbol{\Sigma}^{-1}$ is

$$\boldsymbol{\Sigma}^{-1} | \mathbf{X}, \mathbf{Y} \sim W(\mathbf{V}^{-1}, p, N - q).$$

Thus, inferences about variances (diagonal elements of $\boldsymbol{\Sigma}$) can be made from inverted gamma distributions (the marginal distributions of the diagonal elements of $\boldsymbol{\Sigma}$).

5.3.4 Predictive Density

In the multivariate regression model, suppose that we have a new independent variable observation $\mathbf{w}: (q \times 1)$ and that we wish to predict the corresponding dependent variable observation, say, $\mathbf{z}: (p \times 1)$. Extending the univariate regression approach to the multivariate case, it is not hard to show [see, e.g., Press (1982, p. 420)] that the density of the predictive distribution is given by

$$p\Big(\underset{(p \times 1)}{\mathbf{z}} | \mathbf{X}, \mathbf{Y}\Big) \propto \frac{1}{\{\nu + (\mathbf{z} - \hat{\mathbf{B}}\mathbf{w})'\mathbf{H}(\mathbf{z} - \hat{\mathbf{B}}\mathbf{w})\}^{(\nu + p)/2}},$$

where

$$\nu = N - q - p + 1, \qquad \mathbf{H} \equiv \frac{\nu \mathbf{V}^{-1}}{1 + \mathbf{w}'\mathbf{D}^{-1}\mathbf{w}}, \qquad \mathbf{D} \equiv \sum_1^N \mathbf{x}_j \mathbf{x}_j',$$

where $\mathbf{x}_j$ denotes the jth independent variable observation.

5.4 SUMMARY

In this chapter we have shown how to carry out Bayesian inferences in the classical univariate and multivariate multiple linear regression models, assuming normal disturbances with spherical covariance structures. We found posterior distributions for the regression coefficients, and we also found predictive densities for new observations.

EXERCISES

5.1 Consider the univariate multiple regression model

$$y_i|\mathbf{x}_i = \beta_0 + \beta_1 x_{i1} + \cdots + \beta_p x_{ip} + u_i, \qquad i = 1, \ldots, n,$$

where y_i denotes yield for a crop in year i; x_{ij} denotes the observed value of explanatory variable j during year i; u_i denotes a disturbance term in year i; $\boldsymbol{\beta} \equiv (\beta_i): (p+1) \times 1$ is a vector of coefficients to be estimated; and $\mathbf{x}_i \equiv (x_{ij})$ is a vector of observed explanatory variables during year i. Assume the u_i's are independent $N(0, \sigma^2)$. Introduce a vague prior for $(\boldsymbol{\beta}, \sigma^2)$.

(a) Find the joint posterior density for $(\boldsymbol{\beta}, \sigma^2)$.
(b) Find the marginal posterior density for $\boldsymbol{\beta}$.
(c) Find the marginal posterior density for σ^2.
(d) Find the marginal posterior density for β_1.
(e) Find the predictive density for a new observation y^*, based upon a new observation vector $\mathbf{x}^* \equiv (x_j^*)$.

5.2 Consider the simple univariate regression model of Section 5.1. Introduce a natural conjugate prior for $(\beta_1, \beta_2, \sigma^2)$.

(a) Find the joint posterior density for $(\beta_1, \beta_2, \sigma^2)$.
(b) Find the marginal posterior density for (β_1, β_2).
(c) Find the marginal posterior density for σ^2.
(d) Find the marginal posterior density for β_2.
(e) Find the predictive density for a new observation y^*, based upon a new observation x^*.

5.3 For the simple univariate regression model of Section 5.1, suppose that instead of assuming the u_i's are independent $N(0, \sigma^2)$, we assume that they are independent and that $u_i \sim N(0, \sigma_i^2)$, where $\sigma_i^2 = \sigma^2 x_i$ (heteroscedasticity). Answer parts (a)–(e) of Exercise 5.2, using the interpretation of σ_i^2 in this problem.

5.4 For the simple univariate regression model of Section 5.1, suppose that instead of assuming the u_i's are independent, we assume serially correlated errors with $\text{corr}(u_i, u_j) = \rho$, $i \neq j$, $\text{var}(u_i) = \sigma^2$, $0 < \rho < 1$. Adopt a vague prior for $(\beta_1, \beta_2, \sigma^2, \rho)$ by taking $(\beta_1, \beta_2, \sigma^2)$ as vague in the usual way, assume ρ is a priori independent of $(\beta_1, \beta_2, \sigma^2)$, and assume a uniform prior for ρ on the unit interval.

(a) Find the joint posterior density for $(\beta_1, \beta_2, \sigma^2, \rho)$.

(b) Find the marginal posterior density for ρ.

(c) Find the marginal posterior density for β_2.

(d) Find the marginal posterior density for σ^2.

(e) Find the predictive density for a new observation y^* based upon a new observation x^*.

5.5 For the model in Exercise 5.2, give a highest posterior density credibility interval for β_2, of 95% probability.

5.6 For the model in Exercise 5.2, give:

(a) the modal Bayesian estimator for β_2;

(b) the minimum-risk Bayesian estimator with respect to a quadratic loss function for β_2;

(c) the minimum-risk Bayesian estimator with respect to an absolute error loss function for β_2.

5.7 Suppose $\mathbf{X}$: $p \times 1$ and $\mathscr{L}(\mathbf{X}|\boldsymbol{\theta}) = N(\boldsymbol{\theta}, \mathbf{I})$. Give a natural conjugate prior family for $\boldsymbol{\theta}$.

5.8 Suppose $\mathbf{X}$: $p \times 1$ and $\mathscr{L}(\mathbf{X}|\boldsymbol{\Lambda}) = N(\mathbf{0}, \boldsymbol{\Sigma})$ for $\boldsymbol{\Lambda} \equiv \boldsymbol{\Sigma}^{-1}$. Give a natural conjugate prior family for $\boldsymbol{\Lambda}$. (Assume $\boldsymbol{\Sigma}$ is a positive definite matrix.)

5.9 Suppose $\mathbf{X}$: $p \times 1$ and $\mathscr{L}(\mathbf{X}|\boldsymbol{\theta}, \boldsymbol{\Sigma}) = N(\boldsymbol{\theta}, \boldsymbol{\Sigma})$ for $\boldsymbol{\Sigma}$ a positive definite matrix. Find a natural conjugate prior family for $(\boldsymbol{\theta}, \boldsymbol{\Lambda})$ by noting that $p(\boldsymbol{\theta}, \boldsymbol{\Lambda}) = p_1(\boldsymbol{\theta}|\boldsymbol{\Lambda})p_2(\boldsymbol{\Lambda})$ and $\boldsymbol{\Lambda} \equiv \boldsymbol{\Sigma}^{-1}$ and by assuming that a priori, $\text{var}(\boldsymbol{\theta}|\boldsymbol{\Sigma}) \propto \boldsymbol{\Sigma}$.

5.10* Use the example in Section 5.1.1, involving crop yield in year i, y_i, and quantity of fertilizer used in year i, x_i, and test the hypothesis (using Bayesian procedures) that H: $\beta_2 = 3$, versus the alternative hypothesis that A: $\beta_2 \neq 3$ at the 5% level of credibility (see Section 2.3.2):

(a) Use the Jeffreys procedure for testing, and assume $p(H) = p(A) = 1/2$ (*Hint:* see also Jeffreys, 1961).

(b) Use the Lindley testing procedure (Lindley, 1965) of rejecting H if the 95% credibility interval for $(\beta_2 - 3)$ does not include the origin.

5.11 Do Exercise 5.1 assuming a g-prior for $(\boldsymbol{\beta}, \sigma^2)$ instead of the vague prior (*Hint:* see Section 2.7.3).

*Exercise 5.10 requires reference to sources outside of this text. Full reference information can be found in the bibliography at the back of this book.

CHAPTER VI

Bayesian Multivariate Analysis of Variance and Covariance

6.1 INTRODUCTION

In this chapter we develop Bayesian procedures for problems in the multivariate analysis of variance (MANOVA). Our context is one in which it is reasonable to adopt the assumption of exchangeability for the population mean vectors (across populations rather than within populations). We study the one-way classification, and we adopt fixed-effect models (the extension to higher way layouts is very straightforward). We develop posterior distributions for the main and interaction effects, and we find Stein-type shrinkage estimators. We will focus on the MANOVA model for one-way classification. Related work is discussed in Box and Tiao (1973), Broemeling (1985); Lindley and Smith (1972), Press (1980b) and Press and Shigemasu (1985). Problems of the univariate analysis of variance are solved as a special case.

6.2 MANOVA MODEL: ONE-WAY LAYOUT

Adopt the model

$$\underset{(p\times 1)}{\mathbf{y}_{ij}} = \underset{(p\times 1)}{\boldsymbol{\mu}} + \underset{(p\times 1)}{\boldsymbol{\alpha}_i} + \underset{(p\times 1)}{\mathbf{u}_{ij}}, \qquad i = 1,\dots,q; \qquad j = 1,\dots,n_i,$$

where $\mathbf{y}_{ij}$ denotes a p-vector of observations representing the jth replication in the ith population, $\boldsymbol{\mu}$ denotes the grand mean, $\boldsymbol{\alpha}_i$ denotes the main effect due to population i, and $\mathbf{u}_{ij}$ denotes a disturbance term.

Reduction to Regression Format

Let $N = \sum_{i=1}^{n} n_i$, and define

$$\underset{(p\times N)}{\mathbf{Y}'} = [\mathbf{y}_{11},\ldots,\mathbf{y}_{1n_1};\ldots;\mathbf{y}_{q1},\ldots,\mathbf{y}_{qn_q}]$$

$$\underset{(p\times N)}{\mathbf{U}'} = [\mathbf{u}_{11},\ldots,\mathbf{u}_{1n_1};\ldots;\mathbf{u}_{q1},\ldots,\mathbf{u}_{qn_q}]$$

$$\underset{(p\times 1)}{\boldsymbol{\theta}_i} = \boldsymbol{\mu} + \boldsymbol{\alpha}_i,$$

$$\underset{(p\times q)}{\mathbf{B}'} = (\boldsymbol{\theta}_1,\ldots,\boldsymbol{\theta}_q),$$

$$\underset{(N\times q)}{\mathbf{X}} = \begin{pmatrix} \mathbf{e}_{n_1} & & \mathbf{0} \\ & \ddots & \\ \mathbf{0} & & \mathbf{e}_{n_q} \end{pmatrix}, \qquad \underset{(n\times 1)}{\mathbf{e}_n} = \begin{pmatrix} 1 \\ \vdots \\ 1 \end{pmatrix}.$$

The model becomes

$$\underset{(N\times p)}{\mathbf{Y}} = \underset{(N\times q)}{\mathbf{X}} \cdot \underset{(q\times p)}{\mathbf{B}} + \underset{(N\times p)}{\mathbf{U}}.$$

6.3 LIKELIHOOD

Assume the $\mathbf{u}_{ij}$'s are mutually independent and

$$\mathbf{u}_{ij} \sim N(\mathbf{0}, \boldsymbol{\Sigma}), \qquad \boldsymbol{\Sigma} > 0.$$

The likelihood function is then given by

$$p(\mathbf{Y}\mid\mathbf{X},\mathbf{B},\boldsymbol{\Sigma}) = \frac{1}{|\boldsymbol{\Sigma}|^{N/2}} e^{-\operatorname{tr}[\mathbf{V}+(\mathbf{B}-\hat{\mathbf{B}})'\mathbf{S}(\mathbf{B}-\hat{\mathbf{B}})]\boldsymbol{\Sigma}^{-1}/2},$$

where

$$\underset{(q\times p)}{\hat{\mathbf{B}}} = (\mathbf{X}'\mathbf{X})^{-1}\mathbf{X}'\mathbf{Y}, \qquad \underset{(q\times q)}{\mathbf{S}} = \mathbf{X}'\mathbf{X},$$

$$\underset{(p\times p)}{\mathbf{V}} = (\mathbf{Y} - \mathbf{X}\hat{\mathbf{B}})'(\mathbf{Y} - \mathbf{X}\hat{\mathbf{B}}).$$

6.4 PRIORS

1. Assume that **B** and Σ are independent, that is,

$$p(\mathbf{B}, \Sigma) = p(\mathbf{B})p(\Sigma).$$

2. Assume that Σ follows an "inverted Wishart distribution" (and therefore, Σ^{-1} follows a Wishart distribution), so that for some hyperparameters $(\nu, \mathbf{H})$ we have

$$p(\Sigma) \propto \frac{1}{|\Sigma|^{\nu/2}} e^{-\operatorname{tr} \Sigma^{-1}\mathbf{H}/2}.$$

3. For $\mathbf{B}' \equiv (\theta, \dots, \theta_q)$, take the θ_i to be i.i.d. (and therefore exchangeable) and normally distributed. Note that if the correlations between pairs of θ_i vectors are small, exchangeable θ_i's will be approximately independent (when they are $N(\xi, \Phi)$). Note that we are not assuming vague prior distributions. Thus,

$$p(\mathbf{B}) = \prod_{i=1}^{q} p(\theta_i),$$

$$(\theta_i | \xi, \Phi) \sim N(\xi, \Phi), \qquad \Phi > 0,$$

so that

$$p(\mathbf{B}|\xi, \Phi) \propto \frac{1}{|\Phi|^{q/2}} e^{-\Sigma_1^q (\theta_i - \xi)'\Phi^{-1}(\theta_i - \xi)/2}.$$

6.5 PRACTICAL IMPLICATIONS OF THE EXCHANGEABILITY ASSUMPTION IN THE MANOVA PROBLEM

Assume that the prior distribution for the mean vectors of the populations in Section 6.2 is invariant under reordering of the populations (the populations are exchangeable; see Section 2.9.2). Under nonexchangeability, the matrix of coefficients $\underset{(p\times q)}{\mathbf{B}'} \equiv (\theta_1, \dots, \theta_q)$ has (pq) elements. If they are assumed to be jointly normally distributed, there would be a total (mean + variances and covariances) of $[pq + pq(pq+1)/2]$ hyperparameters to assess. For example, if $q = 3$ and $p = 5$ without exchangeability, there would be 135 parameters to assess.

Under exchangeability we must assess only $\boldsymbol{\xi}: (p \times 1)$ and $\boldsymbol{\Phi}: (p \times p)$ for a total of $[p + p(p+1)/2]$ hyperparameters. For $p = 5$, this implies 20 assessments instead of 135.

Other Implications

Without exchangeability, as the number of populations, q, increases, the number of hyperparameters that must be assessed increases beyond the 135 in this example; with exchangeability we never have to assess more than 20 (for $p = 5$), regardless of the size of q.

Without exchangeability, the elements of $\mathbf{B}'$ are taken to be correlated; under exchangeability the columns of $\mathbf{B}'$ are assumed to be independent. This takes advantage of the basic population structure in the problem (by assuming that the means of the populations are independent, a priori). Define

$$\underset{(pq\times 1)}{\boldsymbol{\theta}} = \begin{pmatrix} \boldsymbol{\theta}_1 \\ \vdots \\ \boldsymbol{\theta}_q \end{pmatrix}, \qquad \underset{(pq\times 1)}{\boldsymbol{\theta}^*} = \mathbf{e}_q \otimes \boldsymbol{\xi} = \begin{pmatrix} \boldsymbol{\xi} \\ \vdots \\ \boldsymbol{\xi} \end{pmatrix}$$

and note that

$$\sum_1^q (\boldsymbol{\theta}_i - \boldsymbol{\xi})'\boldsymbol{\Phi}^{-1}(\boldsymbol{\theta}_i - \boldsymbol{\xi}) = (\boldsymbol{\theta} - \boldsymbol{\theta}^*)'\left(\mathbf{I}_q \otimes \boldsymbol{\Phi}^{-1}\right)(\boldsymbol{\theta} - \boldsymbol{\theta}^*).$$

We are using the notation $\otimes$ for the "direct product" of two matrices. It is defined by the following, for

$$\underset{(p\times q)}{\mathbf{A}} \equiv (a_{ij}), \quad \underset{(r\times s)}{\mathbf{B}} \equiv (b_{ij}): \mathbf{A} \otimes \mathbf{B} = \begin{pmatrix} a_{11}\mathbf{B} & a_{12}\mathbf{B} & \cdots & a_{1q}\mathbf{B} \\ \vdots & \vdots & & \vdots \\ a_{p1}\mathbf{B} & a_{p2}\mathbf{B} & \cdots & a_{pq}\mathbf{B} \end{pmatrix}$$

$$= \mathbf{C}: pr \times qs.$$

6.6 POSTERIOR

6.6.1 Joint Posterior

$$p(\mathbf{B}, \boldsymbol{\Sigma} | \boldsymbol{\xi}, \boldsymbol{\Phi}, \mathbf{X}, \mathbf{Y}) \propto \frac{1}{|\boldsymbol{\Phi}|^{q/2}} e^{-\operatorname{tr}(\mathbf{B}-\mathbf{B}^*)(\mathbf{B}-\mathbf{B}^*)'\boldsymbol{\Phi}^{-1}/2}$$

$$\times \frac{1}{|\boldsymbol{\Sigma}|^{(N+\nu)/2}} e^{-\operatorname{tr}[(\mathbf{V}+\mathbf{H})+(\mathbf{B}-\hat{\mathbf{B}})'\mathbf{S}(\mathbf{B}-\hat{\mathbf{B}})]\boldsymbol{\Sigma}^{-1}/2}$$

or, after completing the square in **B**,

$$p(\mathbf{B}, \boldsymbol{\Sigma} | \boldsymbol{\xi}, \boldsymbol{\Phi}, \mathbf{X}, \mathbf{Y}) \propto \frac{1}{|\boldsymbol{\Sigma}|^{(N+\nu)/2}} e^{-\{\operatorname{tr}(\mathbf{H}+\mathbf{V})\boldsymbol{\Sigma}^{-1} + C(\boldsymbol{\Sigma})\}/2}$$

$$\times e^{-(\boldsymbol{\theta}-\bar{\boldsymbol{\theta}})'[(\mathbf{I}_q \otimes \boldsymbol{\Phi}^{-1}) + (\mathbf{S}\otimes\boldsymbol{\Sigma}^{-1})](\boldsymbol{\theta}-\bar{\boldsymbol{\theta}})/2}$$

where

$$C(\boldsymbol{\Sigma}) \equiv (\hat{\boldsymbol{\theta}} - \boldsymbol{\theta}^*)'\Big\{\big(\mathbf{I}_q \otimes \boldsymbol{\Phi}^{-1}\big)\big[\big(\mathbf{I}_q \otimes \boldsymbol{\Phi}^{-1}\big) + (\mathbf{S} \otimes \boldsymbol{\Sigma}^{-1})\big]^{-1}(\mathbf{S} \otimes \boldsymbol{\Sigma}^{-1})\Big\}$$

$$\times (\hat{\boldsymbol{\theta}} - \boldsymbol{\theta}^*)$$

and

$$\bar{\boldsymbol{\theta}} = \big[\big(\mathbf{I}_q \otimes \boldsymbol{\Phi}^{-1}\big) + (\mathbf{S} \otimes \boldsymbol{\Sigma}^{-1})\big]^{-1}\big[\big(\mathbf{I}_q \otimes \boldsymbol{\Phi}^{-1}\big)\boldsymbol{\theta}^* + (\mathbf{S} \otimes \boldsymbol{\Sigma}^{-1})\hat{\boldsymbol{\theta}}\big].$$

6.6.2 Conditional Posterior

Thus, conditionally,

$$(\boldsymbol{\theta} | \boldsymbol{\Sigma}, \boldsymbol{\xi}, \boldsymbol{\Phi}, \mathbf{X}, \mathbf{Y}) \sim N\Big\{\bar{\boldsymbol{\theta}}, \big[\big(\mathbf{I}_q \otimes \boldsymbol{\Phi}^{-1}\big) + (\mathbf{S} \otimes \boldsymbol{\Sigma}^{-1})\big]^{-1}\Big\}.$$

Note: If

$$\mathbf{Q} \equiv \big[(\mathbf{I} \otimes \boldsymbol{\Phi}^{-1}) + (\mathbf{S} \otimes \boldsymbol{\Sigma}^{-1})\big]^{-1}(\mathbf{I} \otimes \boldsymbol{\Phi}^{-1})$$

so

$$(\mathbf{I} - \mathbf{Q}) \equiv \big[(\mathbf{I} \otimes \boldsymbol{\Phi}^{-1}) + (\mathbf{S} \otimes \boldsymbol{\Sigma}^{-1})\big]^{-1}(\mathbf{S} \otimes \boldsymbol{\Sigma}^{-1}),$$

then

$$\bar{\boldsymbol{\theta}} = \mathbf{Q}\boldsymbol{\theta}^* + (\mathbf{I} - \mathbf{Q})\hat{\boldsymbol{\theta}}.$$ □

This a matrix weighted average, Stein-type estimator (see Berger, 1980; Efron and Morris, 1973; and James and Stein, 1960).

6.6.3 Marginal Posterior

Integrating $\boldsymbol{\Sigma}$ out of the joint posterior for $(\mathbf{B}, \boldsymbol{\Sigma})$ (by noting that $\boldsymbol{\Sigma}^{-1}$ in the integral has a Wishart density) gives

$$p(\mathbf{B} | \mathbf{X}, \mathbf{Y}) \propto \frac{e^{-\Sigma_1^q(\boldsymbol{\theta}_i - \boldsymbol{\xi})'\boldsymbol{\Phi}^{-1}(\boldsymbol{\theta}_i - \boldsymbol{\xi})/2}}{|\boldsymbol{\Phi}|^{q/2}|(\mathbf{V} + \mathbf{H}) + (\mathbf{B} - \hat{\mathbf{B}})'\mathbf{S}(\mathbf{B} - \hat{\mathbf{B}})|^{\delta/2}}$$

where

$$\delta = N + \nu - p - 1.$$

Equivalently, since $\boldsymbol{\Phi}$ is assumed to be a known hyperparameter, we have

$$p(\mathbf{B}|\mathbf{X},\mathbf{Y}) \propto \frac{e^{-(\boldsymbol{\theta}-\boldsymbol{\theta}^*)'(\mathbf{I}_q \otimes \boldsymbol{\Phi}^{-1})(\boldsymbol{\theta}-\boldsymbol{\theta}^*)/2}}{|(\mathbf{V}+\mathbf{H}) + (\mathbf{B}-\hat{\mathbf{B}})'\mathbf{S}(\mathbf{B}-\hat{\mathbf{B}})|^{\delta/2}}.$$

We have shown [see Press (1982, p. 255), where $\mathbf{B}$ plays the role of $\mathbf{B}'$ here] that in large samples, it is approximately true that

$$(\boldsymbol{\theta}|\mathbf{X},\mathbf{Y}) \sim N(\bar{\boldsymbol{\theta}}, \mathbf{M}),$$

where

$$\bar{\boldsymbol{\theta}} = \mathbf{K}\boldsymbol{\theta}^* + (\mathbf{I}-\mathbf{K})\hat{\boldsymbol{\theta}},$$

$$\mathbf{K} = \mathbf{M}\left(\mathbf{I}_q \otimes \boldsymbol{\Phi}^{-1}\right),$$

$$\mathbf{M} = \left\{\left(\mathbf{I}_q \otimes \boldsymbol{\Phi}^{-1}\right) + (\mathbf{S} \otimes \hat{\boldsymbol{\Sigma}}^{-1})\right\}^{-1},$$

$$\hat{\boldsymbol{\Sigma}} = \frac{\mathbf{V}+\mathbf{H}}{N}.$$

Thus, in large samples, the posterior mean, $\bar{\boldsymbol{\theta}}$, is a matrix weighted average of the prior mean and the MLE (i.e., a Stein-type estimator).

6.6.4 Example: Balanced Design

Take $n_1 = n_2 = \cdots = n_q = n$. Then,

$$\mathbf{S} = \mathbf{X}'\mathbf{X} = n\mathbf{I}_q$$

and

$$\bar{\boldsymbol{\theta}} = \mathbf{K}\boldsymbol{\theta}^* + (\mathbf{I}-\mathbf{K})\hat{\boldsymbol{\theta}},$$

where

$$\mathbf{K} = \mathbf{I}_q \otimes \left\{\mathbf{I}_p + n\boldsymbol{\Phi}\hat{\boldsymbol{\Sigma}}^{-1}\right\}^{-1}.$$

So

$$\bar{\boldsymbol{\theta}}_i = \left(\mathbf{I}_p + n\boldsymbol{\Phi}\hat{\boldsymbol{\Sigma}}^{-1}\right)^{-1}\left\{\boldsymbol{\xi} + n\boldsymbol{\Phi}\hat{\boldsymbol{\Sigma}}^{-1}\hat{\boldsymbol{\theta}}_i\right\}, \qquad i = 1,\ldots,q.$$

That is, the posterior mean of the ith group is a matrix weighted average of the prior mean $\boldsymbol{\xi}$ and the MLE in large samples. Also, for large n and $\boldsymbol{\Omega} \equiv [\boldsymbol{\Phi}^{-1} + n\hat{\boldsymbol{\Sigma}}^{-1}]^{-1}$, we have

$$(\boldsymbol{\theta}_i | \text{Data}) \sim N(\bar{\boldsymbol{\theta}}_i, \boldsymbol{\Omega}).$$

Case of p = 1

Take $\Phi = \phi^2$, $\Sigma = \sigma^2$. Then, in terms of precisions, we have

$$\bar{\theta}_i = \left[\frac{(\phi^2)^{-1}}{(\phi^2)^{-1} + \left(\frac{\hat{\sigma}^2}{n}\right)^{-1}}\right]\xi + \left[\frac{\left(\frac{\hat{\sigma}^2}{n}\right)^{-1}}{(\phi^2)^{-1} + \left(\frac{\hat{\sigma}^2}{n}\right)^{-1}}\right]\hat{\theta}_i.$$

Interval Estimation. Since in large samples $\boldsymbol{\theta}_i = \boldsymbol{\mu} + \boldsymbol{\alpha}_i$, $i = 1, \ldots, q$, is normally distributed, interval estimates of $\boldsymbol{\theta}_i$ are readily effected. In cases where $\boldsymbol{\xi}$ or $\boldsymbol{\Phi}$ are known only imperfectly, we use matrix **T**-distributions for interval estimation (i.e., we use the marginals of matrix **T**-distributions, which are multivariate **t**-distributions).

6.7 EXAMPLE: TEST SCORES

Suppose $\mathbf{y}_{ij}$ denotes the bivariate test scores (verbal and quantitative) of the jth student in a sample of students in state i taking a national test (say $i = 1, 2, 3$; $j = 1, \ldots, 100$; $p = 2$; $q = 3$). We want to compare mean scores of students across three states.

Model

$$\underset{(2\times 1)}{\mathbf{y}_{ij}} = \boldsymbol{\mu} + \boldsymbol{\alpha}_i + \mathbf{u}_{ij} \equiv \boldsymbol{\theta}_i + \mathbf{u}_{ij}, \quad \text{with } \boldsymbol{\theta}_i \equiv \boldsymbol{\mu} + \boldsymbol{\alpha}_i,$$

and

$$\mathbf{u}_{ij} \sim N(\mathbf{0}, \boldsymbol{\Sigma}), \qquad \text{i.i.d., } \boldsymbol{\Sigma} > 0.$$

We propose to carry out a one-way-classification fixed-effects MANOVA, from a Bayesian point of view. Define $\underset{(2\times 3)}{\mathbf{B}'} = (\boldsymbol{\theta}_1, \boldsymbol{\theta}_2, \boldsymbol{\theta}_3)$ and assume $p(\mathbf{B}, \boldsymbol{\Sigma}) = p(\mathbf{B})p(\boldsymbol{\Sigma})$,

$$\boldsymbol{\Sigma} \sim W^{-1}(\mathbf{H}, 2, \nu), \qquad \mathbf{H} = (h_{ij}) > 0.$$

Assume also

$$\boldsymbol{\theta}_i \sim N(\boldsymbol{\xi}, \boldsymbol{\Phi}), \qquad \text{i.i.d. (exchangeability)}.$$

Suppose we assess

$$\boldsymbol{\xi} = \begin{pmatrix} 425 \\ 425 \end{pmatrix}, \qquad \boldsymbol{\Phi} = \begin{pmatrix} 2500 & 1250 \\ 1250 & 2500 \end{pmatrix} \equiv \begin{pmatrix} \sigma^2 & \sigma^2\rho \\ \sigma^2\rho & \sigma^2 \end{pmatrix}$$

($\sigma = 50$, $\rho = 1/2$); we also assess

$$E(\boldsymbol{\Sigma}) = \frac{\mathbf{H}}{\nu - 2p - 2} = \frac{\mathbf{H}}{\nu - 6} = \begin{pmatrix} 10{,}000 & 5{,}000 \\ 5{,}000 & 10{,}000 \end{pmatrix}.$$

Based upon informed guesses about the variances of the elements of $\boldsymbol{\Sigma}$, we take $\nu = 100$. Then,

$$\mathbf{H} = 10^4 \begin{pmatrix} 94 & 47 \\ 47 & 94 \end{pmatrix}.$$

We have assessed nine parameters.

Since the samples are large we have, approximately, the posterior distribution

$$(\boldsymbol{\theta} | \text{Data}) \sim N(\bar{\boldsymbol{\theta}}, \mathbf{M}).$$

Suppose that the MLE's are given by

$$\hat{\boldsymbol{\theta}}_1 = \begin{pmatrix} 460 \\ 450 \end{pmatrix}, \qquad \hat{\boldsymbol{\theta}}_2 = \begin{pmatrix} 420 \\ 430 \end{pmatrix}, \qquad \hat{\boldsymbol{\theta}}_3 = \begin{pmatrix} 390 \\ 460 \end{pmatrix}$$

and that the sum-of-squares-of-residuals matrix is

$$\mathbf{V} = 10^5 \begin{pmatrix} 14.60 & 7.30 \\ 7.30 & 11.60 \end{pmatrix}.$$

Then,

$$\hat{\boldsymbol{\Sigma}} = \frac{\mathbf{V} + \mathbf{H}}{N} = 10^3 \begin{pmatrix} 8 & 4 \\ 4 & 7 \end{pmatrix},$$

where

$$N = \sum_1^3 n_i, \qquad n_i = 100, \qquad \nu = 100, \qquad q = 3.$$

Assume the design is balanced. Then, from the result in Section 6.6.3, we have

$$(\boldsymbol{\theta}_i | \text{Data}) \sim N(\bar{\boldsymbol{\theta}}_i, \boldsymbol{\Omega}),$$

where

$$\bar{\boldsymbol{\theta}}_1 = \begin{pmatrix} 459 \\ 449 \end{pmatrix}, \quad \bar{\boldsymbol{\theta}}_2 = \begin{pmatrix} 420 \\ 430 \end{pmatrix}, \quad \bar{\boldsymbol{\theta}}_3 = \begin{pmatrix} 391 \\ 459 \end{pmatrix}, \quad \boldsymbol{\Omega} = \begin{pmatrix} 78 & 39 \\ 39 & 68 \end{pmatrix}.$$

Note that each element of $\bar{\boldsymbol{\theta}}_i$ lies between the corresponding elements of $\boldsymbol{\xi}$ and $\hat{\boldsymbol{\theta}}_i$, $i = 1, 2, 3$ (we have deleted the final decimals).

Contrasts

$$(\boldsymbol{\theta}_1 - \boldsymbol{\theta}_2) | \text{Data} \sim N(\bar{\boldsymbol{\theta}}_i - \bar{\boldsymbol{\theta}}_j, 2\boldsymbol{\Omega}).$$

For example,

$$(\boldsymbol{\theta}_1 - \boldsymbol{\theta}_2) | \text{Data} \sim N\left[\begin{pmatrix} 39 \\ 19 \end{pmatrix}, \begin{pmatrix} 156, & 78 \\ 78, & 136 \end{pmatrix}\right].$$

Thus,

$$P\left\{\begin{matrix}\text{Mean verbal score in state 1} \\ \text{exceeds mean verbal score in state 2}\end{matrix} \middle| \text{Data}\right\} = 99.91\%,$$

and we can compute any other credibility intervals as easily.

6.8 POSTERIOR DISTRIBUTIONS OF EFFECTS

Since

$$\boldsymbol{\theta}_i = \boldsymbol{\mu} + \boldsymbol{\alpha}_i, \qquad i = 1, \ldots, q,$$

the $\boldsymbol{\mu}$ and $\boldsymbol{\alpha}_i$ effects are confounded, as is well known in MANOVA.

If we adopt the usual identifying constraint,

$$\frac{1}{q} \sum_1^q \boldsymbol{\alpha}_i = 0,$$

we see that

$$(1) \quad \boldsymbol{\mu} = \frac{1}{q} \sum_1^q \boldsymbol{\theta}_i,$$

and

$$(2) \quad \boldsymbol{\alpha}_i = \boldsymbol{\theta}_i - \frac{1}{q}\sum_1^q \boldsymbol{\theta}_i.$$

In the case of a balanced design (e.g., in large samples), since we know that

$$(\boldsymbol{\theta}_i | \text{Data}) \sim N(\bar{\boldsymbol{\theta}}_i, \boldsymbol{\Omega}),$$

the posterior distributions of the effects are therefore

$$(1) \quad (\boldsymbol{\mu} | \text{Data}) \sim N\left(\frac{1}{q}\sum_1^q \bar{\boldsymbol{\theta}}_i, \frac{\boldsymbol{\Omega}}{q}\right)$$

and

$$(2) \quad (\boldsymbol{\alpha}_i | \text{Data}) \sim N\left(\bar{\boldsymbol{\theta}}_i - \frac{1}{q}\sum_1^q \bar{\boldsymbol{\theta}}_i, \boldsymbol{\Omega}\left(1 - \frac{1}{q}\right)\right).$$

Alternatively, we could adopt the assumption of exchangeability of the $\boldsymbol{\alpha}_i$'s instead of using the constraints. That is, instead of adopting the normal prior on $\boldsymbol{\theta}_i$, as in Section 6.4, we could assume that $\boldsymbol{\mu} + \boldsymbol{\alpha}_i$ are independent, a priori, and that each is normally distributed. Their distributions are then combined, and analysis proceeds as before. (See Exercise 6.4.)

6.9 SUMMARY

This chapter has presented methods for making Bayesian inferences in the multivariate analysis of variance and covariance by adopting the assumption of exchangeability of the distribution of populations means.

EXERCISES

6.1 Provide conditions under which a set of exchangeable random variables might be treated approximately as if they were independent.

6.2* Go to the original sources and explain the notion of "partial exchangeability." [*Hint:* See Diaconis and Freedman (1980, 1984).]

*Asterisked exercises require reference to sources outside of this text. Full reference information can be found in the bibliography at the back of this book.

6.3* What can be said about de Finetti's theorem for finitely exchangeable events? [*Hint;* see Diaconis (1977).]

6.4 Following the suggestion at the end of Section 6.8, assume that $\boldsymbol{\mu} + \boldsymbol{\alpha}_i$ are independent a priori, and assume that the $\boldsymbol{\alpha}_i$'s are exchangeable and that $\boldsymbol{\alpha}_i \sim N(\boldsymbol{\xi}^*, \boldsymbol{\Phi}^*)$ for all i.

(a) Find the prior density for $\boldsymbol{\theta}_i$.

(b) Find the prior density for **B**.

(c) Find the joint posterior density for $(\mathbf{B}, \boldsymbol{\Sigma})$.

6.5* Adopt the two-way layout

$$\underset{(p\times1)}{\mathbf{y}_{ijk}} = \underset{(p\times1)}{\boldsymbol{\mu}} + \underset{(p\times1)}{\boldsymbol{\alpha}_i} + \underset{(p\times1)}{\boldsymbol{\beta}_j} + \underset{(p\times1)}{\boldsymbol{\gamma}_{ij}} + \underset{(p\times1)}{\mathbf{u}_{ijk}},$$

where $i = 1, \ldots, I$; $j = 1, \ldots, J$; $k = 1, \ldots, K$; and the main effects, interaction effects, and disturbances are interpreted as in the conventional ANOVA model. Explain how you would generalize the results of this chapter for the one-way layout to the two-way layout. [*Hint:* See Press and Shigemasu (1985).]

6.6 How would you generalize the results of this chapter for the one-way layout to a one-way layout with covariates (MANOCOVA)?

6.7 Explain why the Bayesian estimators in the example involving test scores in Section 6.7 are so close to the MLE's.

CHAPTER VII

Bayesian Inference in Classification and Discrimination

7.1 INTRODUCTION

The problem of classifying an observation vector into one of several populations from which it must have come is well known. We focus here on the predictive Bayesian approach that began with the work of de Finetti (1937), and then Geisser (1964, 1966, 1967). [For a general discussion of classification, as well as a comparison of Bayesian with frequentist methods, see, for example, Press (1982).] A summary is given below.

There are K populations $\pi_1, \ldots, \pi_K$, where

$$\pi_i \equiv N(\boldsymbol{\theta}_i, \boldsymbol{\Sigma}_i), \qquad i = 1, \ldots, K.$$

$(\boldsymbol{\theta}_i, \boldsymbol{\Sigma}_i)$ are assumed to be unknown. $\mathbf{z}: (p \times 1)$ is an observation known to come from one of the K populations, but we don't know which one. So the observation is to be correctly classified.

We have available "training samples"; that is,

$$\begin{array}{llll} \mathbf{x}_1^{(1)}, \ldots, \mathbf{x}_{N_1}^{(1)} & & \text{are i.i.d. from} & \pi_1 \\ \vdots & \vdots & \vdots & \vdots \\ \mathbf{x}_1^{(K)}, \ldots, \mathbf{x}_{n_K}^{(K)} & & \text{are i.i.d. from} & \pi_K. \end{array}$$

Approach

1. Introduce a prior on $(\boldsymbol{\theta}_i, \boldsymbol{\Sigma}_i)$, and develop the posterior.
2. Find the predictive distribution for $(\mathbf{z} | \pi_i)$.
3. Find the posterior probability $P\{\mathbf{z} \in \pi_i | \mathbf{z}\}$.

REMARK: When the parameters of the π_i's are known (which is unusual in practice), Bayesian and frequentist procedures coincide. In this case, the indicated procedure is: Classify $\mathbf{z}$ into that population for which a linear or quadratic function of $\mathbf{z}$ is maximum (linear, if the Σ_i's are the same; quadratic, otherwise). The predictive Bayesian classification approach was extended to classifying $\mathbf{z}$ using its neighbors (contextual classification), and to classifying $\mathbf{z}$ when all data are spatially correlated (see Klein and Press, 1987, 1988).

Preliminaries. The MLE's are

$$\hat{\boldsymbol{\theta}}_i = \bar{\mathbf{x}}_i = \frac{1}{N_i} \sum_{j=1}^{N_i} \mathbf{x}_j^{(i)}$$

and

$$\hat{\boldsymbol{\Sigma}}_i = \frac{\mathbf{V}_i}{N_i} = \frac{1}{N_i} \sum_{j=1}^{N_i} \left(\mathbf{x}_j^{(i)} - \bar{\mathbf{x}}_i\right)\left(\mathbf{x}_j^{(i)} - \bar{\mathbf{x}}_i\right)'.$$

The unbiased estimator of $\boldsymbol{\Sigma}_i$ is

$$\hat{\boldsymbol{\Sigma}}_i^* = \frac{\mathbf{V}_i}{N_i - 1}.$$

Denote the "precision matrix" by

$$\boldsymbol{\Lambda}_i \equiv \boldsymbol{\Sigma}_i^{-1}, \qquad \boldsymbol{\Sigma}_i > 0,$$

and parameterize the problem in terms of $\boldsymbol{\Lambda}_i$. Then, if $p(\boldsymbol{\theta}_i, \boldsymbol{\Lambda}_i)$ denotes the prior density, the posterior density is given by

$$p(\boldsymbol{\theta}_i, \boldsymbol{\Lambda}_i | \bar{\mathbf{x}}_i, \mathbf{V}_i, \pi_i) \propto p(\bar{\mathbf{x}}_i, \mathbf{V}_i | \boldsymbol{\theta}_i, \boldsymbol{\Lambda}_i, \pi_i) p(\boldsymbol{\theta}_i, \boldsymbol{\Lambda}_i).$$

7.2 LIKELIHOOD FUNCTION

Since

$$(\bar{\mathbf{x}}_i | \boldsymbol{\theta}_i) \sim N\left(\boldsymbol{\theta}_i, \frac{\boldsymbol{\Sigma}_i}{N_i}\right)$$

and

$$(\mathbf{V}_i | \boldsymbol{\Lambda}_i, \pi_i) \sim W(\boldsymbol{\Lambda}_i^{-1}, p, n_i) \qquad \text{(i.e., a Wishart distribution)},$$

where

$$n_i \equiv N_i - 1, \quad \text{and} \quad p \leq n_i,$$

it follows that

$$p(\bar{\mathbf{x}}_i|\boldsymbol{\theta}_i, \boldsymbol{\Lambda}_i, \pi_i) \propto |\boldsymbol{\Lambda}_i|^{1/2} e^{-(N_i/2)(\bar{\mathbf{x}}_i - \boldsymbol{\theta}_i)'\boldsymbol{\Lambda}_i(\bar{\mathbf{x}}_i - \boldsymbol{\theta}_i)}$$

and

$$p(\mathbf{V}_i|\boldsymbol{\Lambda}_i, \pi_i) \propto |\mathbf{V}_i|^{(n_i - p - 1)/2} |\boldsymbol{\Lambda}_i|^{n_i/2} e^{-\operatorname{tr} \boldsymbol{\Lambda}_i \mathbf{V}_i/2}.$$

The likelihood function can therefore be expressed as

$$\begin{aligned} p(\mathbf{x}_i, \mathbf{V}_i|\boldsymbol{\theta}_i, \boldsymbol{\Lambda}_i, \pi_i) &= p(\bar{\mathbf{x}}_i|\boldsymbol{\theta}_i, \boldsymbol{\Lambda}_i, \pi_i) p(\mathbf{V}_i|\boldsymbol{\Lambda}_i, \pi_i) \\ &\propto |\mathbf{V}_i|^{(n_i - p - 1)/2} |\boldsymbol{\Lambda}_i|^{N_i/2} e^{-\operatorname{tr} \boldsymbol{\Lambda}_i[\mathbf{V}_i + N_i(\bar{\mathbf{x}}_i - \boldsymbol{\theta}_i)(\bar{\mathbf{x}}_i - \boldsymbol{\theta}_i)']/2}. \end{aligned}$$

7.3 PRIOR DENSITY

We can assess a complete prior distribution for $(\boldsymbol{\theta}_i, \boldsymbol{\Lambda}_i)$ at this point (we could use natural conjugate priors as well). Alternatively, we use a vague prior (see Sections 2.7.2 and 5.3.1):

$$p(\boldsymbol{\theta}_i, \boldsymbol{\Lambda}_i) \propto \frac{1}{|\boldsymbol{\Lambda}_i|^{(p+1)/2}}.$$

7.4 POSTERIOR DENSITY

The posterior density becomes

$$p(\boldsymbol{\theta}_i, \boldsymbol{\Lambda}_i|\bar{\mathbf{x}}_i, \mathbf{V}_i, \pi_i) \propto |\boldsymbol{\Lambda}_i|^{(N_i - p - 1)/2} e^{-\operatorname{tr} \boldsymbol{\Lambda}_i[\mathbf{V}_i + N_i(\bar{\mathbf{x}}_i - \boldsymbol{\theta}_i)(\bar{\mathbf{x}}_i - \boldsymbol{\theta}_i)']/2}.$$

7.5 PREDICTIVE DENSITY

Lemma. The predictive density of $(\mathbf{z}|\bar{\mathbf{x}}_i, \mathbf{V}_i, \pi_i)$ is multivariate **t** and is given by

$$p(\mathbf{z}|\bar{\mathbf{x}}_i, \mathbf{V}_i, \pi_i) \propto \frac{1}{\left[1 + \dfrac{N_i}{N_i^2 - 1}(\mathbf{z} - \bar{\mathbf{x}}_i)'\mathbf{S}_i^{-1}(\mathbf{z} - \bar{\mathbf{x}}_i)\right]^{N_i/2}},$$

where

$$\mathbf{S}_i \equiv \frac{\mathbf{V}_i}{N_i - 1}.$$

Proof (Outline)

$$p(\mathbf{z}|\bar{\mathbf{x}}_i, \mathbf{V}_i, \pi_i) \equiv \int\int p(\mathbf{z}|\boldsymbol{\theta}_i, \boldsymbol{\Lambda}_i, \pi_i) p(\boldsymbol{\theta}_i, \boldsymbol{\Lambda}_i|\bar{\mathbf{x}}_i, \mathbf{V}_i, \pi_i)\, d\boldsymbol{\theta}_i\, d\boldsymbol{\Lambda}_i.$$

But $(\mathbf{z}|\boldsymbol{\theta}_i, \boldsymbol{\Lambda}_i, \pi_i) \sim N(\boldsymbol{\theta}_i, \boldsymbol{\Lambda}_i^{-1})$. Substituting the normal and posterior densities into the integral gives

$$p(\mathbf{z}|\bar{\mathbf{x}}_i, \mathbf{V}_i, \pi_i) \propto \int\int |\boldsymbol{\Lambda}_i|^{(N_i - p)/2} e^{-\operatorname{tr} \boldsymbol{\Lambda}_i \mathbf{A}/2}\, d\boldsymbol{\theta}_i\, d\boldsymbol{\Lambda}_i,$$

where

$$\mathbf{A} \equiv \mathbf{V}_i + N_i(\bar{\mathbf{x}}_i - \boldsymbol{\theta}_i)(\bar{\mathbf{x}}_i - \boldsymbol{\theta}_i)' + (\mathbf{z} - \boldsymbol{\theta}_i)(\mathbf{z} - \boldsymbol{\theta}_i)';$$

note that $\mathbf{A}$ does not depend upon $\boldsymbol{\Lambda}_i$.

Integrating with respect to $\boldsymbol{\Lambda}_i$ by using the Wishart density properties [see e.g., Press (1982)] gives

$$p(\mathbf{z}|\bar{\mathbf{x}}_i, \mathbf{V}_i, \pi_i) \propto \int \frac{d\theta_i}{|\mathbf{A}|^{(N_i+1)/2}}.$$

Note that $\mathbf{A}$ contains two terms that are quadratic in $\boldsymbol{\theta}_i$. Completing the square in $\boldsymbol{\theta}_i$, and simplifying with matrix algebra, gives

$$p(\mathbf{z}|\bar{\mathbf{x}}_i, \mathbf{V}_i, \pi_i) \propto \frac{1}{|\mathbf{F}_i|^{(N_i+1)/2}} \int \frac{d\boldsymbol{\theta}_i}{\left[(N_j + 1)^{-1} + (\boldsymbol{\theta}_i - \boldsymbol{\alpha}_i)'\mathbf{F}_i^{-1}(\boldsymbol{\theta}_i - \boldsymbol{\alpha}_i)\right]^{(N_i+1)/2}},$$

where

$$\alpha_i = \left(\frac{N_i}{N_i + 1}\right)\bar{\mathbf{x}}_i + \left(\frac{1}{N_i + 1}\right)\mathbf{z}$$

and

$$\mathbf{F}_i = (\mathbf{V}_i + N_i\bar{\mathbf{x}}_i\bar{\mathbf{x}}_i' + \mathbf{z}\mathbf{z}') - \frac{(N_i\bar{\mathbf{x}}_i + \mathbf{z})(N_i\bar{\mathbf{x}}_i + \mathbf{z})'}{N_i + 1}.$$

The last integration is readily carried out by noting that the integrand is the kernel of a multivariate Student's t-density. We find

$$p(\mathbf{z}|\bar{\mathbf{x}}_i, \mathbf{V}_i, \pi_i) \propto \frac{1}{|\mathbf{F}_i|^{N_i/2}}.$$

Simplifying gives the required result. □

7.6 POSTERIOR CLASSIFICATION PROBABILITY

Let q_i = prior probability that $\mathbf{z}$ comes from π_i; that is,

$$q_i = P\{\mathbf{z} \in \pi_i\}.$$

We preassign this value. By Bayes' theorem,

$$P\{\mathbf{z} \in \pi_i|\mathbf{z}\} = \frac{P\{\mathbf{z} \in \pi_i\}\, p(\mathbf{z}|\bar{\mathbf{x}}_i, \mathbf{V}_i, \pi_i)}{\sum_{j=1}^{K} p(\mathbf{z}|\bar{\mathbf{x}}_j, \mathbf{V}_j, \pi_j) P\{\mathbf{z} \in \pi_j\}}.$$

The posterior odds is given by

$$\frac{P\{\mathbf{z} \subset \pi_i|\mathbf{z}\}}{P\{\mathbf{z} \in \pi_j|\mathbf{z}\}} = \left(\frac{q_i}{q_j}\right) \frac{p(\mathbf{z}|\bar{\mathbf{x}}_i, \mathbf{V}_i, \pi_i)}{p(\mathbf{z}|\bar{\mathbf{x}}_j, \mathbf{V}_j, \pi_j)}$$

$$= L_{ij} \frac{\left\{1 + \left(\dfrac{N_j}{N_j^2 - 1}\right)(\mathbf{z} - \bar{\mathbf{x}}_j)'\mathbf{S}_j^{-1}(\mathbf{z} - \bar{\mathbf{x}}_j)\right\}^{N_j/2}}{\left\{1 + \left(\dfrac{N_i}{N_i^2 - 1}\right)(\mathbf{z} - \bar{\mathbf{x}}_i)'\mathbf{S}_i^{-1}(\mathbf{z} - \bar{\mathbf{x}}_i)\right\}^{N_i/2}},$$

where

$$L_{ij} = \left(\frac{q_i}{q_j}\right) \frac{|(N_j - 1)\mathbf{S}_j|^{1/2}}{|(N_i - 1)\mathbf{S}_i|^{1/2}} \left[\frac{\Gamma\left(\dfrac{N_i}{2}\right)\Gamma\left(\dfrac{N_j - p}{2}\right)}{\Gamma\left(\dfrac{N_j}{2}\right)\Gamma\left(\dfrac{N_i - p}{2}\right)}\right] \left[\frac{N_i(N_j + 1)}{N_j(N_i + 1)}\right]^{p/2},$$

for all $i, j = 1, \ldots, K$..

REMARKS

1. For Bayesian classification with unknown parameters, we need only compute the ratios of pairs of Student's **t**-densities to see which of the two populations is more likely (the last equation in Section 7.6).
2. We need not make any assumption about equality of covariance matrices.
3. Sizes of the training samples need not be large for the procedure to be valid.
4. If natural conjugate families of priors are used, the same structural form results, except the parameters are different.
5. If $N_i = N_j$, L_{ij} reduces to

$$L_{ij} = \left(\frac{q_i}{q_j}\right)\sqrt{\frac{|\mathbf{S}_j|}{|\mathbf{S}_i|}}.$$

7.7 EXAMPLE: TWO POPULATIONS

Suppose there are two normal populations ($K = 2$) that are each two-dimensional ($p = 2$), with unknown parameters. We want to classify an observation $\mathbf{z}$. Suppose the training samples give

$$\bar{\mathbf{x}}_1 = \begin{pmatrix} 1 \\ 1 \end{pmatrix}, \qquad \bar{\mathbf{x}}_2 = \begin{pmatrix} 0 \\ 0 \end{pmatrix},$$

$$\mathbf{S}_1 = \begin{pmatrix} 1 & 1/2 \\ 1/2 & 1 \end{pmatrix}, \qquad \mathbf{S}_2 = \begin{pmatrix} 1 & 1/4 \\ 1/4 & 1/2 \end{pmatrix},$$

based upon $N_1 = N_2 = N = 10$ observations. Since $|\mathbf{S}_1| = 3/4$ and $|\mathbf{S}_2| = 7/16$, if the prior probabilities are equal ($q_1 = q_2 = 1/2$) we obtain

$$L_{12} = .76.$$

So

$$\frac{p(\mathbf{z} \in \pi_1|\mathbf{z})}{p(\mathbf{z} \in \pi_2|\mathbf{z})} = (.76)\frac{\left\{1 + \frac{10}{99}(\mathbf{z} - \bar{\mathbf{x}}_2)'S_2^{-1}(\mathbf{z} - \bar{\mathbf{x}}_2)\right\}^5}{\left\{1 + \frac{10}{99}(\mathbf{z} - \bar{\mathbf{x}}_1)'S_1^{-1}(\mathbf{z} - \bar{\mathbf{x}}_1)\right\}^5}.$$

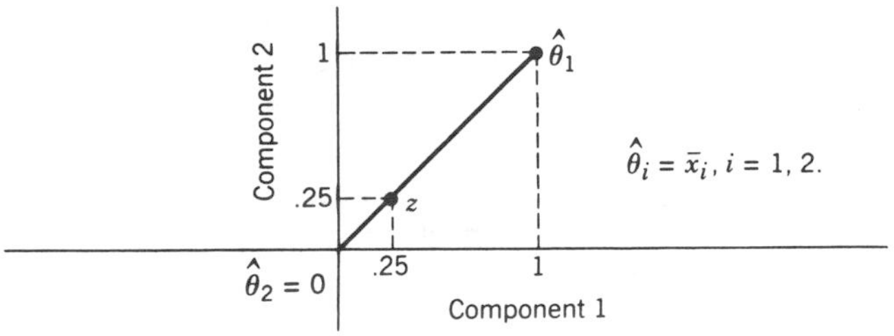

Figure 7.1. Location of z relative to the centers of two populations.

Suppose $\mathbf{z} = \begin{pmatrix} 1/4 \\ 1/4 \end{pmatrix}$. We then find

$$\frac{p(\mathbf{z} \in \pi_1 | \mathbf{z})}{p(\mathbf{z} \in \pi_2 | \mathbf{z})} = .57.$$

Although the prior odds ratio on π_1 compared to π_2 was $1:1$, the posterior odds are down to .57 to 1; that is, it is almost two to one in favor of π_2. (See Figure 7.1 for a pictorial presentation of the location of $\mathbf{z}$ relative to the centers of the two populations.)

7.8 SECOND GUESSING UNDECIDED RESPONDENTS: AN APPLICATION

In this section we apply Bayesian classification procedures to a problem in sample surveys, namely, that of estimating category probabilities when we have some respondents who are difficult to classify.

7.8.1 Problem

In sample surveys and opinion polls involving sensitive questions, some subjects respond in the category "Don't know," "No opinion," or "Undecided," to avoid stating their true beliefs (not because they truly don't know because they are uninformed). How can these subjects be correctly classified by the survey analyst?

Solution

We assume there is a main question of interest, and we also assume that there are some subsidiary questions. In order to classify the "undecided" respondents, we propose (1) drawing upon the information obtained from subjects who respond unambiguously on both the main question and subsidiary questions and (2) taking advantage of the correlations among the responses to these two types of questions.

Suppose there are n_i subjects who respond unambiguously in category i of the main question, $i = 1, \ldots, M$. Suppose further that there are m subjects who are "undecided" on the main question but who answer unambiguously on the subsidiary questions. The joint responses of the subjects who respond unambiguously on the main question follow a multinomial distribution with category probabilities $q_1, \ldots, q_M$. We are interested in estimating the category probabilities q_i, taking into account the "undecided" group. We can give Bayesian estimators of the q_i's, given all of the observed data.

The Bayesian estimators have been shown (see Press and Yang, 1974) to be (assuming vague priors) given by

$$\hat{q}_i = \frac{n_i + 1}{m + \sum_{j=1}^{M}(n_j + 1)} \sum_{j=1}^{m} \frac{h(\mathbf{z}^{(j)}|\pi_i)}{\sum_{t=1}^{M}(n_t + 1) h(\mathbf{z}^{(j)}|\pi_t)},$$

where: $\hat{q}_i$ denotes the mean of the posterior distribution of q_i, given all of the observed data; $h(\mathbf{z}^{(j)}|\pi_i)$ denotes the marginal predictive density of the response to the subsidiary questions, $\mathbf{z}^{(j)}$, for the jth "undecided" respondent, if he/she had responded in category i on the main question (i.e., he/she was in population π_i); n_i is the number of subjects who respond unambiguously in category i of the main question; and m is the number of "undecideds."

The subsidiary questions may have categorical, continuous, or mixed categorical and continuous responses. In the special case where the subsidiary questions have categorical responses, we have

$$h(\mathbf{z}^{(j)}|\pi_i) = h(\mathbf{z}^{(j)} = \mathbf{u}_k|\pi_i) = \frac{x(k|i) + 1}{n_i + S},$$

where: $\mathbf{u}_k$ denotes an $S \times 1$ vector with a "one" in the kth place and zeros elsewhere; $x(k|i)$ denotes the number of respondents who answered unambiguously in category i on the main question and cell k of the subsidiary group; and S denotes the number of cells in the contingency table of responses to the subsidiary question set.

The posterior variance of q_i is given by

$$\sigma_{q_i}^2 = E(q_i^2|\mathbf{Z}, \mathbf{N}) - [E(q_i|\mathbf{Z}, \mathbf{N})]^2,$$

where

$$E(q_i|\mathbf{Z},\mathbf{N}) \equiv \hat{q}_i,$$

and

$$AE\left(q_i^2|\mathbf{Z},\mathbf{N}\right) = (n_i + 1)(n_i + 2) + 2(n_i + 2)H + W,$$

where

$$H \equiv \sum_{j=1}^{m} P\{\pi_i|\mathbf{z}^{(j)}\} = \sum_{j=1}^{m}\left[\frac{(n_i+1)h\left(\mathbf{z}^{(j)}|\pi_i\right)}{\sum_{k=1}^{M}(n_k+1)h\left(\mathbf{z}^{(j)}|\pi_k\right)}\right],$$

$$\mathbf{N} \equiv (n_1, \dots, n_{M-1}), \qquad \mathbf{Z} \equiv (z^{(1)}, \dots, z^{(m)}),$$

$$W \equiv 2\sum_{j_1=1}^{m-1}\sum_{j_2=j_1+1}^{m} P\{\pi_i, \pi_i|\mathbf{z}^{(j_1)}, \mathbf{z}^{(j_2)}\},$$

$$A \equiv \left[m + \sum_{i=1}^{M}(n_i+1)\right]\left[(m+1) + \sum_{i=1}^{M}(n_i+1)\right],$$

$$P\{\pi_i, \pi_i|\mathbf{z}^{(j_1)}, \mathbf{z}^{(j_2)}\} = C(n_i+1)(n_i+2)h\left(\mathbf{z}^{(j_1)}|\pi_i\right)h\left(\mathbf{z}^{(j_2)}|\pi_i\right),$$

$$C^{-1} = \sum_{i=1}^{M}(n_i+2)(n_i+1)h\left(\mathbf{z}^{(j_1)}|\pi_i\right)h\left(\mathbf{z}^{(j_2)}|\pi_i\right)$$

$$+ \sum_{\substack{i=1 \\ (i \neq j)}}^{M}\sum_{j=1}^{M}\left((n_i+1)(n_j+1)h\left(\mathbf{z}^{(j_1)}|\pi_i\right)h\left(\mathbf{z}^{(j_2)}|\pi_j\right)\right).$$

Proofs of these summary results can be found in Press and Yang (1974).

7.8.2 Illustration

Suppose there are 100 respondents to a sensitive question, such that 20 people respond in each of three possible, unambiguous categories, and 40 are "undecided." Suppose further that there is just one subsidiary question (with four possible response categories). The "decided" group on the main question responded as shown in Table 7.1. Thus, 17 subjects responded in category 1 of the subsidiary question, given they responded in category 1 on the main question.

Table 7.1 Responses of the "Decided" Group

Main Question Category	Subsidiary Question Category 1	2	3	4	Total
1	17	1	1	1	20 (33%)
2	5	5	5	5	20 (33%)
3	1	1	1	17	20 (33%)

For the "undecided" group, suppose 25 respond in subsidiary question category 1, and also suppose that 5 respond in each of the other three categories.

Since there were an equal number of subjects who responded in each of the three categories of the main question (20 subjects in each), we have

$$h\left(z^{(j)} = u_k | \pi_i\right) = \frac{x(k|i) + 1}{n_i + S} = \frac{x(k|i) + 1}{20 + 4} \quad \text{for } i = 1, \ldots, 3.$$

The values of $h(\cdot | \cdot)$ are arrayed in Table 7.2. Bayesian classification yields the results shown in Table 7.3. Thus, if the undecided respondents were to be ignored, we would be making substantial errors, $(\hat{q}_i - \tilde{q}_i)/\tilde{q}_i$: (.062/.333, −.004/.333, −.057/.333) = (18.6%, −1.2%, −17.1%), for point estimates of the category probabilities; in addition, we have entire distributions for the category probabilities, so we can readily evaluate dispersion of our estimates and tail probabilities.

Table 7.2 Values of $h(\cdot | \cdot)$

Main Question Category	Subsidiary Question Category 1	2	3	4
1	$\frac{17+1}{24}$	$\frac{1+1}{24}$	$\frac{1+1}{24}$	$\frac{1+1}{24}$
2	$\frac{5+1}{24}$	$\frac{5+1}{24}$	$\frac{5+1}{24}$	$\frac{5+1}{24}$
3	$\frac{1+1}{24}$	$\frac{1+1}{24}$	$\frac{1+1}{24}$	$\frac{17+1}{24}$

Table 7.3 Results of Bayesian Classification

Main Question Category (i)	$\tilde{q}_i$ Ignoring "Undecideds"	$\hat{q}_i$ Second Guessing the "Undecideds"	σ_{q_i}	90% Credibility Interval
1	.333	.395	.062	(.30, .50)
2	.333	.329	.057	(.24, .43)
3	.333	.276	.050	(.20, .36)

7.9 SUMMARY

We have shown that by using Bayesian procedures for classification: (1) the methods are exact (not approximate or large sample); (2) it is unnecessary to assume equal covariance matrices; and (3) results can be used to improve accuracy and understanding in opinion polling.

EXERCISES

7.1 Suppose there are two normal populations that are each two-dimensional with unknown parameters. We want to classify an observation z. Suppose the training samples give

$$\bar{\mathbf{x}}_1 = (2,3)', \qquad \bar{\mathbf{x}}_2 = (3,1)',$$

$$\mathbf{S}_1 = \begin{pmatrix} 2 & 1 \\ 1 & 4 \end{pmatrix}, \qquad \mathbf{S}_2 = \begin{pmatrix} 3 & 2 \\ 2 & 4 \end{pmatrix},$$

based upon $N_1 = N_2 = N = 10$ observations. Let the observed vector be given by

$$\mathbf{z} = (2, 1.5)'.$$

If the prior odds are equal, classify $\mathbf{z}$ into one of the two populations.

7.2 Suppose there are three bivariate normal populations with training sample statistics as in Exercise 7.1, but, in addition, we know $N_1 = N_2 = N_3 = N = 10$, and

$$\bar{\mathbf{x}}_3 = (0,0), \qquad \mathbf{S}_3 = \begin{pmatrix} 3 & 1 \\ 1 & 1 \end{pmatrix}.$$

Assuming the prior odds for all three populations are the same, classify

$$\mathbf{z} = (2,2)'$$

into one of the three populations.

7.3 For the classification problem in Section 7.1 adopt a natural conjugate prior family for $(\boldsymbol{\theta}_i, \boldsymbol{\Lambda}_i)$, instead of the vague prior family used there, and find:

(a) the joint posterior density of $\boldsymbol{\theta}_i$, $\boldsymbol{\Lambda}_i$, given $(\bar{\mathbf{x}}_i, \mathbf{V}_i, \pi_i)$;

(b) the predictive density of $(\mathbf{z}|\bar{\mathbf{x}}_i, \mathbf{V}_i, \pi_i)$;

(c) the posterior classification probability for classifying $\mathbf{z}$ into π_i, given $\mathbf{z}$, that is $P\{\mathbf{z} \in \pi_i|\mathbf{z}\}$.

7.4 Explain the classification concept behind the notion of "second guessing undecided respondents," as discussed in Section 7.8.

7.5 In what types of situations would we be likely to encounter large fractions of "undecided" respondents, in the sense presented in Section 7.8?

7.6* While it is a general principle that it is better to have more information than less, why might it be better to have very few subsidiary questions in the undecided respondents problem? (*Hint:* see Press and Yang, 1974.)

*Exercise 7.6 requires reference to sources outside of this text. Full reference information can be found in the bibliography at the back of this book.

CHAPTER VIII

A Case Study in Applied Bayesian Inference

8.1 INTRODUCTION

The theory of Bayesian inference we summarized in earlier chapters has been applied in many fields, including biology (e.g., medical diagnosis), business (e.g., managerial decision making), economics (e.g., econometrics), engineering (e.g., Kalman filtering; control theory), law (e.g., paternity testing; see Section 2.11), linguistics (e.g., disputed authorship), and psychology (e.g., psychometric testing; growth modeling), just to name a few. Such applications may be found, for example, in selected articles appearing in collections in Aykac and Brumat (1976), Bernardo et al. (1980, 1985), Fienberg and Zellner (1975), Grayson (1960), Kanji (1983), and Zellner (1971).

To illustrate how Bayesian inference is carried out in a life-size real problem involving large data sets with real prior information, we borrow from a Bayesian analysis of disputed authorship of The Federalist papers. This analysis is described in the book by Mosteller and Wallace (1964); see also Mosteller and Wallace (1984), which is a second edition of the book with an annotated table of contents and an additional chapter of references to recent research on authorship problems. We present below a very brief summary of the problem, along with its Bayesian resolution. The reader should refer to Mosteller and Wallace (1984) for additional details of this interesting application of Bayesian methods. Subsequently, when we wish to cite this 1984 reference, we will refer to "M & W," for shorthand convenience.

8.2 THE FEDERALIST PAPERS

The Federalist papers are a collection of 77 essays (plus eight written later) that appeared in 1787–1788, to persuade citizens of the State of New York to ratify the constitution proposed for the recently established new country, the United States of America. They were short essays, only about 900–3500 words in length. They appeared originally in newspapers but now appear in collected form as a book. We will be concerned with the 12 essays (of the original 77) whose authorship has been in dispute. Was the author of each of these 12 Alexander Hamilton or James Madison (and no one else)? In addition to the 12 essays with disputed authorship, there were 43 essays known to have been written by Hamilton, 14 known to have been written by Madison, and five known to have been written by John Jay. An additional three are known to have been coauthored by Hamilton and Madison. We accept the "known" attributions given here. (See M & W regarding the uncertainties in these attributions.) The problem of statistical linguistics facing us is, who wrote the 12 remaining essays?

M & W decided to use the papers of known authorship (and some other known authorship works) to select certain simple words used by Hamilton and Madison in their writings, to differentiate their writing styles. After empirically testing the "discrimination strength" of many possible candidates for words that could be used for discrimination, they finally fixed upon 30 words. Computers were used to carry out word counts for the 30 final words counted in works with known authorship. Usage rates (rates of usage of a given word per thousand words of text) were established for each author, for each word. Each discriminator word was treated independently of every other discriminator word (M & W made studies of dependence of several words and found such dependence to be relatively unimportant). To establish usage rates, 94,000 words of text known to have been written by Hamilton were used, and 114,000 words of text known to have been written by Madison were used (works other than Madison's Federalist papers were used as a supplement).

8.3 LIKELIHOOD FUNCTION

M & W adopted the negative binomial distribution for their likelihood function (they also tried a Poisson likelihood, but it did not fit the observed data as well). The usual representation of the negative binomial distribution is a probability mass function given by

$$f(x|\kappa, p) = \frac{\Gamma(x+\kappa)}{x!\Gamma(\kappa)} p^{\kappa}(1-p)^{x}, \qquad x = 0, 1, \ldots,$$

where: $\Gamma(t)$ denotes the gamma function of argument t; x denotes the number of failures before the κth success, in independent Bernoulli trials; and p denotes the probability of success in a single trial. If θ denotes the odds of a failure in a single trial, then $\theta \equiv (1-p)/p$. Reparameterizing in terms of θ gives

$$f(x|\kappa, \theta) = \frac{\Gamma(x+\kappa)\theta^{x}}{x!\Gamma(\kappa)(1+\theta)^{x+\kappa}}, \qquad x = 0, 1, \ldots . \tag{8.1}$$

M & W give this likelihood function the interpretation that: x denotes the number of occurrences of a word, $\theta \equiv w\delta$, where w denotes the text length (number of words in a text), δ denotes the "non-Poissonness rate" (per unit length), and θ denotes the odds of a particular discriminating word not appearing. An additional parameterization M & W give is $\mu = \kappa\delta$, where μ denotes the mean rate of appearance of a particular discriminating word in a text (mean usage rate per unit length). The negative binomial likelihoods for all 30 of the final words were multiplied to form the joint likelihood for the 30 words, for each of the two authors.

The formalism for making a determination of which of the two authors wrote each of the 12 papers of disputed authorship was based upon posterior odds, computed from Bayes' theorem. This procedure is outlined in the next section.

8.4 POSTERIOR ODDS FOR DISPUTED AUTHORSHIP

Let $\mathbf{x}_j \equiv (x_{ij})$, where x_{ij} denotes the number of occurrences of word i, of origin j, where $j = 1$ if Hamilton wrote the paper and $j = 2$ if Madison wrote the paper; $i = 1, \ldots, 30$. Let $\boldsymbol{\kappa}_j \equiv (\kappa_{ij})$ and $\boldsymbol{\theta}_j \equiv (\theta_{ij})$ denote the corresponding vectors of parameters. The likelihood function for a paper of known origin j is

$$L(\mathbf{x}_j|\boldsymbol{\kappa}_j, \boldsymbol{\theta}_j) = \prod_{i=1}^{30} \left[\frac{\Gamma(x_{ij}+\kappa_{ij})\theta_{ij}^{x_{ij}}}{x_{ij}!\Gamma(\kappa_{ij})(1+\theta_{ij})^{x_{ij}+\kappa_{ij}}} \right]. \tag{8.2}$$

The posterior probability density of the parameters is then, by Bayes' theorem,

$$h_j(\boldsymbol{\kappa}_j, \boldsymbol{\theta}_j | \mathbf{x}_j) \propto g_j(\boldsymbol{\kappa}_j, \boldsymbol{\theta}_j) L(\mathbf{x}_j | \boldsymbol{\kappa}, \theta_j), \tag{8.3}$$

where $g_j(\boldsymbol{\kappa}_j, \boldsymbol{\theta}_j)$ denotes the prior probability density for $(\boldsymbol{\kappa}_j, \boldsymbol{\theta}_j)$.

We now follow the classification approach developed in Chapter V. Let $\mathbf{y} \equiv (y_i)$ denote a vector of discriminator words from a paper of unknown origin; $i = 1, \ldots, 30$. Let q_j denote the prior probability that $\mathbf{y}$ comes from a paper written by author j. By Bayes' theorem,

$$\begin{aligned} &P\{\mathbf{y} \text{ comes from paper of author } j | \mathbf{y}\} \\ &\quad \equiv P\{\text{author } j | \mathbf{y}\} \\ &\quad \propto P\{\, y \text{ comes from paper of author } j\,\} L(\mathbf{y} | \mathbf{x}_j), \end{aligned}$$

or

$$P\{\text{author } j | \mathbf{y}\} \propto q_j L(\mathbf{y} | \mathbf{x}_j).$$

Equivalently, the posterior classification odds ratio is

$$\frac{P\{\text{Hamilton} | \mathbf{y}\}}{P\{\text{Madison} | \mathbf{y}\}} = \left(\frac{q_1}{q_2}\right)\left[\frac{L(\mathbf{y} | \mathbf{x}_1)}{L(\mathbf{y} | \mathbf{x}_2)}\right].$$

Taking logs, for convenience, if the posterior log odds ratio is defined as

$$(\text{LO})_2 \equiv \log\left[\frac{P\{\text{Hamilton} | \mathbf{y}\}}{P\{\text{Madison} | \mathbf{y}\}}\right],$$

and if the prior log odds ratio is defined as

$$(\text{LO})_1 \equiv \log\left(\frac{q_1}{q_2}\right),$$

the posterior log odds ratio becomes

$$(\text{LO})_2 = (\text{LO})_1 + \mathscr{L}, \tag{8.4}$$

where $\mathscr{L}$ denotes the log of the likelihood ratio; i.e,

$$\mathscr{L} \equiv \log \frac{L(\mathbf{y} | \mathbf{x}_1)}{L(\mathbf{y} | \mathbf{x}_2)}. \tag{8.5}$$

If the analysis favors Madison, $(\mathrm{LO})_2$ will be negative, since $P\{\text{Hamilton}\,|\,\mathbf{y}\} < P\{\text{Madison}\,|\,\mathbf{y}\}$; if it favors Hamilton, $(\mathrm{LO})_2$ will be positive.

We note that $L(\mathbf{y}|\mathbf{x}_j)$ denotes the predictive density of the word counts of unknown origin, given the word counts of known origin. This predictive density is obtained from

$$L(\mathbf{y}|\mathbf{x}_j) = \int\int L(\mathbf{y}|\boldsymbol{\kappa}_j, \boldsymbol{\theta}_j) h_j(\boldsymbol{\kappa}_j, \boldsymbol{\theta}_j|\mathbf{x}_j)\, d\boldsymbol{\kappa}_j\, d\boldsymbol{\theta}_j. \tag{8.6}$$

The ratio of predictive densities, $\mathscr{L}$, given in Eqs. (8.5) and (8.6), correspond to Eq. (3), page 100, in M & W. The integrals in Eq. (8.6) were approximated by M & W by using Laplace's method (see also the subsection in Section 3.3.1 entitled "The Tierney–Kadane Approximation," for an improved procedure) to get the modal approximation

$$L(\mathbf{y}|\mathbf{x}_j) \doteq L(\mathbf{y}|\hat{\boldsymbol{\kappa}}_j, \hat{\boldsymbol{\theta}}_j), \tag{8.7}$$

where $(\hat{\boldsymbol{\kappa}}_j, \hat{\boldsymbol{\theta}}_j)$ denotes the mode of the posterior distribution of $(\boldsymbol{\kappa}_j, \boldsymbol{\theta}_j|\mathbf{x}_j)$, and, of course, $L(\mathbf{y}|\hat{\boldsymbol{\kappa}}_j, \hat{\boldsymbol{\theta}}_j)$ denotes the negative binomial likelihood. Thus, to evaluate $(\mathrm{LO})_2$ for each paper, we only need the prior odds ratio, along with the prior distribution of the parameters. M & W take the prior odds to be $1:1$ $(q_1 = q_2)$, so $(\mathrm{LO})_1 = 0$.

8.5 PRIOR DISTRIBUTION OF PARAMETERS

We consider just one word at a time and assume for the moment that (κ_j, θ_j) are scalars. Thus, there are four parameters that index the likelihood for a given discriminator word for the two authors: $(\kappa_1, \theta_1, \kappa_2, \theta_2)$. We transform this quadruple to a new parameterization and adopt our prior on the new quadruple. The transformation is

$$\mathbf{t} \equiv (\kappa_1, \theta_1, \kappa_2, \theta_2) \to \mathbf{t}^*(\omega, \tau, \xi, \eta), \tag{8.8}$$

where

$$\mu_j = \kappa_j \delta_j, \qquad \omega = \mu_1 + \mu_2, \qquad \tau = \frac{\mu_1}{\mu_1 + \mu_2},$$

$$\xi = \log[(1 + \delta_1)(1 + \delta_2)], \qquad \eta = \frac{\log(1 + \delta_1)}{\log[(1 + \delta_1)(1 + \delta_2)]},$$

$$\theta_j = w_j \delta_j, \qquad j = 1, 2,$$

and w_j is assumed known and fixed. We note that $0 \le \tau \le 1$, $0 \le \eta \le 1$, $\omega > 0$, $\xi > 0$. We assume the log posterior odds ratio in Eq. (8.4) is evaluated in terms of the new parameterization $\mathbf{t}^*$ instead of $\mathbf{t}$, so that the posterior analysis is carried out in terms of $\mathbf{t}^*$. Note that μ_j denotes the mean rate of appearance of a particular discriminator word in a paper of authorship j, τ measures the ability to discriminate, and $0 \le \tau \le 1$. When $\tau = \frac{1}{2}$, the word does poorly as a discriminator, since then the word frequencies for the two authors are equal. M & W adopt a symmetric beta prior distribution for τ, with density

$$p(\tau|\omega) = \frac{1}{B(\gamma, \gamma)} \tau^{\gamma-1}(1 - \tau)^{\gamma-1}, \tag{8.9}$$

where $\gamma \equiv \beta_1 + \beta_2\omega$. This assumption on γ permits decreased variability for τ with increased ω. The quantities (β_1, β_2) are hyperparameters (parameters of the prior distribution) and need to be assessed. An empirical study of moments showed M & W that δ_j and μ_j are approximately independent (which was the reason for selecting δ_j as a parameter). Along with M & W we assume (we will use $p(\cdot)$ generically to denote density)

$$p(\omega) \propto \text{constant}. \tag{8.10}$$

Analogously, for (ξ, η) we assume η follows a symmetric beta distribution; but we assume ξ follows a gamma distribution; that is,

$$p(\eta) \propto \eta^{\beta_3-1}(1 - \eta)^{\beta_3-1}, \qquad 0 < \beta_3, \tag{8.11}$$

and

$$p(\xi) \propto \xi^{\beta_5-1} \exp\{-(\beta_5\xi)/\beta_4\}, \qquad 0 < \beta_4, 0 < \beta_5. \tag{8.12}$$

Let $\boldsymbol{\beta} \equiv (\beta_1, \beta_2, \beta_3, \beta_4, \beta_5)$. The elements of $\boldsymbol{\beta}$ are hyperparameters. Once the quintuple of hyperparameters is specified, Eqs. (8.9)–(8.12) define the prior distribution on the parameterization $\mathbf{t}^*$, in Eq. (8.8). M & W carried out many empirical studies to determine appropriate values for the elements of $\boldsymbol{\beta}$. Their final (modal) selection was $\boldsymbol{\beta} \equiv (10, 0, 12, 1.25, 2.0)$. This selection was based upon word distribution in the totality of papers of known authorship. M & W assumed that $\mathbf{t}^*$ values were independent across words. Moreover, they assumed $[(\omega, \tau), \xi, \eta]$ were independent for a given word (actually $(\omega, \tau, \xi, \eta)$ were independent, since they ultimately took

Table 8.1 Final Discriminator Words and Modally Estimated Parameters*

Word	μ_1	μ_2	ω	τ	δ_1	δ_2
1. upon	3.24	.23	3.47	.932	.29	.48
2. also	.32	.67	.99	.327	.13	.14
3. an	5.95	4.58	10.53	.565	.07	.07
4. by	7.32	11.43	18.75	.390	.37	.43
5. of	64.51	57.89	122.40	.527	.26	.28
6. on	3.38	7.75	11.12	.304	.37	.44
7. there	3.20	1.33	4.53	.706	.25	.27
8. this	7.77	6.00	13.77	.564	.25	.23
9. to	40.79	35.21	76.00	.537	.41	.48
10. although	.06	.17	.23	.267	.21	.18
11. both	.52	1.04	1.56	.334	.15	.18
12. enough	.25	.10	.35	.727	.54	.64
13. while	.21	.07	.28	.744	.29	.35
14. whilst	.08	.42	.50	.153	.27	.20
15. always	.58	.20	.78	.742	.13	.13
16. though	.91	.51	1.42	.639	.11	.12
17. commonly	.17	.05	.23	.763	.14	.16
18. consequently	.10	.42	.52	.189	.25	.20
19. considerable(ly)	.37	.17	.54	.684	.13	.15
20. according	.17	.54	.71	.238	.36	.34
21. apt	.27	.08	.35	.770	.13	.16
22. direction	.17	.08	.25	.693	.39	.43
23. innovation(s)	.06	.15	.20	.278	.15	.13
24. language	.08	.18	.26	.316	.11	.10
25. vigor(ous)	.18	.08	.26	.680	.08	.09
26. kind	.69	.17	.86	.799	.29	.27
27. matter(s)	.36	.09	.45	.790	.12	.13
28. particularly	.15	.37	.51	.282	.20	.21
29. probability	.27	.09	.36	.757	.07	.08
30. work(s)	.13	.27	.40	.326	.55	.47

*Note: some of the entries in the table do not exactly satisfy the formulas. This is attributable to rounding.

Table 8.2 Log Posterior Odds Ratios for Disputed Authorship

Disputed Paper Number	Paper Length (thousands of words)	$(LO)_2$ Log Posterior Odds Ratio	Author Attribution
1.	1.6	−13.2	Madison
2.	1.1	−14.3	Madison
3.	1.9	−21.9	Madison
4.	1.8	−16.0	Madison
5.	2.2	−15.8	Madison
6.	2.0	−14.3	Madison
7.	2.0	−5.8	Madison
8.	1.6	−8.7	Madison
9.	2.2	−16.7	Madison
10.	2.1	−18.0	Madison
11.	2.4	−16.5	Madison
12.	3.0	−18.5	Madison

$\beta_2 = 0$). Thus, for a given discriminator word

$$p(\mathbf{t}^*) = p(\omega, \tau, \xi, \eta) = p(\omega)p(\tau)p(\xi)p(\eta).$$

The final discriminator words adopted are given in Table 8.1, along with the posterior modal values estimated for the parameter set $(\mu_1, \mu_2, \omega, \tau, \delta_1, \delta_2)$. Note that (ω, τ) is determined from (μ_1, μ_2).

8.6 FINAL RESULTS

We present the final result of this study of disputed authorship in the form of a table of log posterior odds ratios for each of the 12 disputed authorship papers. These results are given in Table 8.2.

It is clear from Table 8.2 that since all of the $(LO)_2$ are negative, all of the disputed authorship Federalist papers should be attributed to James Madison. Moreover, in the most questionable case (paper #7), the log odds ratio is −5.8, corresponding to an odds ratio of about 330 : 1 in favor of Madison. (The effects of dependencies of words, and uncertainties in the values of hyperparameters, deflated the odds ratio for this paper to 90 : 1.) In the least questionable case (paper #3), the log odds ratio is −21.9, corresponding to an odds ratio of about 3.2 billion to one.

8.7 SUMMARY

In this chapter we have summarized how a detailed, carefully executed, applied study in Bayesian inference in disputed authorship was carried out. There were, of course, many detailed substudies not discussed here which validated (1) the choices of prior distributions selected, (2) the choices of likelihood selected, and (3) the manner in which other important data-analytically based choices were made.

EXERCISES

8.1 Give the density function for the negative binomial distribution, and find its first two central moments.

8.2 Explain why the negative binomial distribution is viewed as a good candidate for modeling the likelihood in a disputed authorship study.

8.3 What would be the effect on the results of the empirical study described in Chapter VIII if the prior odds were established according to the proportions of Federalist papers known to have been written by each of the two principal authors, instead of taking the prior odds as 1 : 1?

8.4 How would you take into account "political content" in a paper of disputed authorship, so that a paper appearing to reflect the political beliefs of a given author would weigh in favor of that author having written the paper?

8.5 How might you improve upon the modal approximation of Eq. (8.7) by using the numerical methods discussed in Chapter III?

8.6* Mosteller and Wallace (1984) also considered using the Poisson distribution as a likelihood function model in the disputed authorship problem. What are the comparative advantages of the negative binomial distribution over the Poisson distribution for this problem (why did M & W select one over the other)?

8.7 Suppose that instead of adopting the prior distribution assumption of Eq. (8.10) we assume $p(\omega) \propto 1/\omega$. Would you anticipate that the final results would be affected very much? Explain.

*Exercise 8.6 requires reference to sources outside of this text. Full reference information can be found in the bibliography at the back of this book.

Appendices

The four appendices include reprints of articles commenting on the life and work of Thomas Bayes, in addition to a reprint of his original essay. The articles are listed below.

Appendix 1 is an article by Hilary Seal entitled "Bayes, Thomas."[1] The article discusses and references articles that followed Bayes' essay. Those articles critique Bayes' original essay.

Appendix 2 is a biographical note on Bayes' original essay, by G. A. Barnard, Imperial College, London.[2]

Appendix 3 is a letter from Mr. Richard Price, a friend of Bayes, to Mr. John Canton, representing the Royal Society of London, submitting Thomas Bayes' original essay for publication after Bayes' death.[3]

Appendix 4 is the original essay of Thomas Bayes, "An Essay Towards Solving a Problem in the Doctrine of Chances," by Reverend Thomas Bayes.[4]

Note: What we normally think of as Bayes' theorem is given (in somewhat obscure form by today's standards) in Proposition 9, Appendix 4.

---

1. reprinted with the permission of the publishers from the *International Encyclopedia of Statistics*, edited by W. H. Kruskal and J. M. Tanur, Volume 1, The Free Press, A Division of Macmillan Publishing Company, Incorporated, New York, 1978.
2. reprinted with the permission of the publishers from *Biometrika*, **45**, Parts 3 and 4, December 1958, 293–295.
3. reprinted with the permission of the publishers from *Biometrika*, **45**, Parts 3 and 4, December 1958, 296–298.
4. reprinted with the permission of the publishers from *Biometrika*, **45**, Parts 3 and 4, December 1958, 298–315. The article originally appeared in *The Philosophical Transactions* (1763), **53**, 370–418, and was reprinted in *Biometrika*.

APPENDIX 1

BAYES, THOMAS

by Hilary L. Seal

Thomas Bayes (1702–1761) was the eldest son of the Reverend Joshua Bayes, one of the first nonconformist ministers to be publicly ordained in England. The younger Bayes spent the last thirty years of his comfortable, celibate life as Presbyterian minister of the meeting house, Mount Sion, in the fashionable town of Tunbridge Wells, Kent. Little is known about his personal history, and there is no record that he communicated with the well-known scientists of his day. Circumstantial evidence suggests that he was educated in literature, languages, and science at Coward's dissenting academy in London (Holland 1962). He was elected a fellow of the Royal Society in 1742, presumably on the basis of two metaphysical tracts he published (one of them anonymously) in 1731 and 1736 (Barnard 1958). The only mathematical work from his pen consists of two articles published posthumously in 1764 by his friend Richard Price, one of the pioneers of social security (Ogborn 1962). The first is a short note, written in the form of an undated letter, on the divergence of the Stirling (de Moivre) series $\ln(z!)$. It has been suggested that Bayes' remark that the use of "a proper number of the first terms of the ... series" will produce an accurate result constitutes the first recognition of the asymptotic behavior of a series expansion (see Deming's remarks in Bayes [1764] 1963). The second article is the famous "An Essay Towards Solving a Problem in the Doctrine of Chances," with Price's preface, footnotes, and appendix (followed, a year later, by a continuation and further development of some of Bayes' results).

The "Problem" posed in the Essay is: "*Given* the number of times in which an unknown event has happened and failed: *Required* the chance that the probability of its happening in a single trial lies somewhere between any two degrees of probability that can be named." A few sentences later Bayes writes: "By *chance* I mean the same as probability" ([1764]. 1963, p. 376).

If the number of successful happenings of the event is p and the failures q, and if the two named "degrees" of probability are b and f, respectively, Proposition 9 of the Essay provides the following answer expressed in terms of areas under the curve $x^p(1-x)^q$:

$$\int_b^f x^p(1-x)^q\,dx \Big/ \int_0^1 x^p(1-x)^q\,dx. \tag{1}$$

This is based on the assumption (Bayes' "Postulate 1") that all values of the unknown probability are equally likely before the observations are made. Bayes indicated the applicability of this postulate in his famous "Scholium": "that the ... rule is the proper one to be used in the case of an event concerning the probability of which we absolutely know nothing antecedently to any trials made concerning it, seems to appear from the following consideration; viz. that concerning such an event I have no reason to think that, in a certain number of trials, it should rather happen any one possible number of times than another" (*ibid*., pp. 392–393).

The remainder of Bayes' Essay and the supplement (half of which was written by Price) consists of attempts to evaluate (1) numerically, (a) by expansion of the integrand and (b) by integration by parts. The results are satisfactory for p and q small but the approximations for large p, q are only of historical interest (Wishart 1927).

Opinions about the intellectual and mathematical ability evidenced by the letter and the essay are extraordinarily diverse. Netto (1908), after outlining Bayes' geometrical proof, agreed with Laplace ([1812] 1820) that it is *ein wenig verwickelt* ("somewhat involved"). Todhunter (1865) thought that the résumé of probability theory that precedes Proposition 9 was "excessively obscure." Molina (in Bayes [1764] 1963, p. xi) said that "Bayes and Price ... can hardly be classed with the great mathematicians that immediately preceded or followed them," and Hogben (1957, p. 133) stated that "the ideas commonly identified with the name of Bayes are largely [Laplace's]."

On the other hand, von Wright (1951, p. 292) found Bayes' Essay "a masterpiece of mathematical elegance and free from ... obscure philosophical pretentions." Barnard (1958, p. 295) wrote that Bayes' "mathematical work ... is of the very highest quality." Fisher ([1956] 1959, p. 8) concurred with these views when he said Bayes' "mathematical contributions ... show him to have been in the first rank of independent thinkers ...".

The subsequent history of mathematicians' and philosophers' extensions and criticisms of Proposition 9—the only statement that can properly be called Bayes' theorem (or rule)—is entertaining and instructive. In his first published article on probability theory, Laplace (1774), without mentioning Bayes, introduced the principle that if p_j is the probability of an observable event resulting from "cause" j ($j = 1, 2, 3, \ldots, n$) then the probability that "cause" j is operative to produce the observed event is

$$p_j \Big/ \sum_{j=1}^{n} p_j. \tag{2}$$

This is Principle III of the first (1812) edition of Laplace's probability text, and it implies that the prior (antecedent, initial) probabilities of each of the "causes" are the same. However, in the second (1814) edition Laplace added a few lines saying that if the "causes" are not equally probable a priori, (2) would become

$$\omega_j p_i \Big/ \sum_{j=1}^{n} \omega_j p_j, \tag{3}$$

where ω_j is the prior probability of cause j and p_j is now the probability of the event, given that "cause" j is operative. He gave no illustrations of this more general formula.

Laplace (1774) applied his new principle (2) to find the probability of drawing m white and n black tickets in a specified order from an urn containing an infinite number of white and black tickets in an unknown ratio and from which p white and q black tickets have already been drawn. His solution, namely,

$$\int_0^1 x^{p+m}(1-x)^{q+n}\,dx \Big/ \int_0^1 x^p(1-x)^q\,dx = \frac{(p+m)!(q+n)!(p+q+1)!}{p!q!(p+q+m+1)!}, \tag{4}$$

was later (1778–1781; 1812, chapter 6) generalized by the bare statement that if all values of x are not equally probable a factor $z(x)$ representing the a priori probability density (*facilité*) of x must appear in both integrands. However, Laplace's own views on the applicability of expressions like (4) were stated in 1778 (1778–1781, p. 264) and agree with those of

Bayes' Scholium: "Lorsqu'on n'a aucune donnée *a priori* sur la possibilité d'un événement, il faut supposer toutes les possibilités, depuis zéro jusqu'à l'unité, également probables. ..." ("When nothing is given a priori as to the probability of an event, one must suppose all probabilities, from zero to one, to be equally likely. ...") Much later Karl Pearson (1924, p. 191) pointed out that Bayes was "considering excess of one variate ... over a second ... as the determining factor of occurrence" and this led naturally to a generalization of the measure in the integrals of (1). Fisher (1956) has even suggested that Bayes himself had this possibility in mind.

Laplace's views about prior probability distributions found qualified acceptance on the Continent (von Kries 1886) but were subjected to strong criticism in England (Boole 1854; Venn 1866; Chrystal 1891; Fisher 1922), where a relative frequency definition of probability was proposed and found incompatible with the uniform prior distribution (for example, E. S. Pearson 1925). However, developments in the theory of inference (Keynes 1921; Ramsey 1923–1928; Jeffreys 1931; de Finetti 1937; Savage 1954; Good 1965) suggest that there are advantages to be gained from a "subjective" or a "logical" definition of probability and this approach gives Bayes' theorem, in its more general form, a central place in inductive procedures (Jeffreys 1939; Raiffa & Schlaifer 1961; Lindley 1965).

BIBLIOGRAPHY

Barnard, G. A. 1958 Thomas Bayes: A Biographical Note. *Biometrika* 45:293–295.

Bayes, Thomas (1764) 1963 *Facsimiles of Two Papers by Bayes*. New York: Hafner → Contains "An Essay Towards Solving a Problem in the Doctrine of Chances, With Richard Price's Foreward and Discussion," with a commentary by Edward C. Molina; and "A Letter on Asymptotic Series From Bayes to John Canton," with a commentary by W. Edwards Deming. Both essays first appeared in Volume 53 of the *Philosophical Transactions*, Royal Society of London, and retain the original pagination.

Boole, George (1854) 1951 *An Investigation of the Laws of Thought, on Which Are Founded the Mathematical Theories of Logic and Probabilities*. New York: Dover.

Chrystal, George 1891 On Some Fundamental Principles in the Theory of Probability. Actuarial Society of Edinburgh, *Transactions* 2:419–439.

[1] de Finetti, Bruno 1937 La prévision: Ses lois logiques, ses sources subjectives. Paris, Université de, Institut Henri Poincaré, *Annales* 7:1-68.

Fisher, R. A. (1922) 1950 On the Mathematical Foundations of Theoretical Statistics. Pages 10.307*a*–10.368 in R. A. Fisher, *Contributions to Mathematical Statistics*. New York: Wiley. → First published in Volume 222 of the *Philosophical Transactions*, Series A, Royal Society of London.

FISHER, R. A. (1956) 1959 *Statistical Methods and Scientific Inference*. 2d ed., rev. New York: Hafner; London: Oliver & Boyd.

GOOD, IRVING J. 1965 *The Estimation of Probabilities: An Essay on Modern Bayesian Methods*. Cambridge, Mass.: M.I.T. Press.

HOGBEN, LANCELOT T. 1957 *Statistical Theory; the Relationship of Probability, Credibility and Error: An Examination of the Contemporary Crisis in Statistical Theory From a Behaviourist Viewpoint*. London: Allen & Unwin.

HOLLAND, J. D. 1962 The Reverend Thomas Bayes, F.R.S. (1702–1761). *Journal of the Royal Statistical Society* Series A 125:451–461.

JEFFREYS, HAROLD (1931) 1957 *Scientific Inference*. 2d ed. Cambridge Univ. Press.

JEFFREYS, HAROLD (1939) 1961 *Theory of Probability*. 3d ed. Oxford: Clarendon.

KEYNES, J. M. (1921) 1952 *A Treatise on Probability*. London: Macmillan. → A paperback edition was published in 1962 by Harper.

KRIES, JOHANNES VON (1886) 1927 *Die Principien der Wahrscheinlichkeitsrechnung: Eine logische Untersuchung*. 2d ed. Tübingen (Germany): Mohr.

LAPLACE, PIERRE S. (1774) 1891 Mémoire sur la probabilité des causes par les événements. Volume 8, pages 27–65 in Pierre S. Laplace, *Oeuvres complètes de Laplace*. Paris: Gauthier-Villars.

LAPLACE, PIERRE S. (1778–1781) 1893 Mémoire sur les probabilités. Volume 9, pages 383–485 in Pierre S. Laplace, *Oeuvres complètes de Laplace*. Paris: Gauthier-Villars.

LAPLACE, PIERRE S. (1812) 1820 *Théorie analytique des probabilités*. 3d ed., rev. Paris: Courcier.

LINDLEY, DENNIS V. 1965 *Introduction to Probability and Statistics From a Bayesian Viewpoint*. 2 vols. Cambridge Univ. Press.

NETTO, E. 1908 Kombinatorik, Wahrscheinlichkeitsrechnung, Reihen-Imaginäres. Volume 4, pages 199–318, in Moritz Cantor (editor), *Vorlesungen über Geschichte der Mathematik*. Leipzig: Teubner.

OGBORN, MAURICE E. 1962 *Equitable Assurances: The Story of Life Assurance in the Experience of The Equitable Life Assurance Society, 1762–1962*. London: Allen & Unwin.

PEARSON, EGON S. 1925 Bayes' Theorem, Examined in the Light of Experimental Sampling. *Biometrika* 17:388–442.

PEARSON, KARL 1924 Note on Bayes' Theorem. *Biometrika* 16:190–193.

PRICE, RICHARD 1765 A Demonstration of the Second Rule in the Essay Towards a Solution of a Problem in the Doctrine of Chances. Royal Society of London, *Philosophical Transactions* 54:296–325. → Reprinted by Johnson in 1965.

RAIFFA, HOWARD; and SCHLAIFER, ROBERT 1961 *Applied Statistical Decision Theory*. Harvard University Graduate School of Business Administration, Studies in Managerial Economics. Boston: The School.

RAMSEY, FRANK P. (1923–1928) 1950 *The Foundations of Mathematics and Other Logical Essays*. New York: Humanities.

SAVAGE, LEONARD J. (1954) 1972 *The Foundations of Statistics*. Rev. ed. New York: Dover. → Includes a new preface.

TODHUNTER, ISAAC (1865) 1949 *A History of the Mathematical Theory of Probability From the Time of Pascal to That of Laplace*. New York: Chelsea.

VENN, JOHN (1866) 1888 *The Logic of Chance: An Essay on the Foundations and Province of the Theory of Probability, With Special Reference to Its Logical Bearings and Its Application to Moral and Social Science*. 3d ed. London: Macmillan.

WISHART, JOHN 1927 On the Approximate Quadrature of Certain Skew Curves, With an Account of the Researches of Thomas Bayes. *Biometrika* 19:1–38.

WRIGHT, GEORG H. VON 1951 *A Treatise on Induction and Probability*. London: Routledge.

POSTSCRIPT

It is difficult to stay current with the extensive literature dealing with methods flowing from Thomas Bayes' original suggestion. It is also difficult to maintain a clear mind in the profusion of discussions about whether to use so-called Bayesian techniques. One may cite the proceedings of the 1970 Symposium on the Foundations of Statistical Inference (1971) and the philosophical treatise by Stegmüller (1973). Two pairs of eminent authors, with long experience in both theory and application of statistics, have adopted very different approaches toward the Bayesian viewpoint: Box and Tiao (1973) and Kempthorne and Folks (1971).

ADDITIONAL BIBLIOGRAPHY

BOX, GEORGE E. P.; and TIAO, GEORGE C. 1973 *Bayesian Inference in Statistical Analysis*. Reading, Mass.: Addison-Wesley.

[1] DE FINETTI, BRUNO (1937) 1964 Foresight: Its Logical Laws, Its Subjective Sources. Pages 93–158 in Henry E. Kyberg, Jr. and Howard E. Smokler, *Studies in Subjective Probability*. New York: Wiley. → First published in French.

KEMPTHORNE, OSCAR; and FOLKS, LEROY 1971 *Probability, Statistics and Data Analysis*, Ames: Iowa State Univ. Press.

STEGMÜLLER, WOLFGANG 1973 *Personelle und statistische Wahrscheinlichkeit*. Volume 2: *Statistisches Schliessen, statistische Begründung, statistische Analyse*. Berlin: Springer.

SYMPOSIUM ON THE FOUNDATIONS OF STATISTICAL INFERENCE, UNIVERSITY OF WATERLOO, *1970* 1971 *Foundations of Statistical Inference: Proceedings*. Edited by V. P. Godambe and D. A. Sprott, Toronto: Holt.

APPENDIX 2

THOMAS BAYES—A BIOGRAPHICAL NOTE

by G. A. Barnard

Bayes's paper, reproduced in the following pages, must rank as one of the most famous memoirs in the history of science and the problem it discusses is still the subject of keen controversy. The intellectual stature of Bayes himself is measured by the fact that it is still of scientific as well as historical interest to know what Bayes had to say on the questions he raised. And yet such are the vagaries of historical records, that almost nothing is known about the personal history of the man. *The Dictionary of National Biography*, compiled at the end of the last century, when the whole theory of probability was in temporary eclipse in England, has an entry devoted to Bayes's father, Joshua Bayes, F.R.S., one of the first six Nonconformist ministers to be publicly ordained as such in England, but it has nothing on his much more distinguished son. Indeed, the note on Thomas Bayes which is to appear in the forthcoming new edition of the *Encyclopedia Britannica* will apparently be the first biographical note on Bayes to appear in a work of general reference since the *Imperial Dictionary of Universal Biography* was published in Glasgow in 1865. And in treatises on the history of mathematics, such as that of Loria (1933) and Cantor (1908), notice is taken of his contributions to probability theory and to mathematical analysis, but biographical details are lacking.

The Reverend Thomas Bayes, F.R.S., author of the first expression in precise, quantitative form of one of the modes of inductive inference, was born in 1702, the eldest son of Ann Bayes and Joshua Bayes, F.R.S. He was educated privately, as was usual with Nonconformists at that time, and from the fact that when Thomas was 12 Bernoulli wrote to Leibniz that 'poor de Moivre' was having to earn a living in London by teaching mathematics, we are tempted to speculate that Bayes may have learned mathematics from one of the founders of the theory of probability. Eventually Thomas was ordained, and began his ministry by helping his father, who was at the time stated, minister of the Presbyterian meeting house in Leather Lane, off Holborn. Later the son went to minister in Tunbridge Wells at the Presbyterian Chapel on Little Mount Sion which had been opened on 1 August 1720. It is not known when Bayes went to Tunbridge Wells, but he was not the first to minister on Little Mount Sion, and he was certainly there in 1731, when he produced a tract entitled 'Divine Benevolence, or an attempt to prove that the Principal End of the Divine Providence and Government is the happiness of His Creatures'. The tract was published by John Noon and copies are in Dr. Williams's library and the British Museum. The following is a quotation:

> [p. 22]: I don't find (I am sorry to say it) any necessary connection between mere intelligence, though ever so great, and the love or approbation of kind and beneficient actions.

Bayes argued that the principal end of the Deity was the happiness of His creatures, in opposition to Balguy and Grove who had, respectively, maintained that the first spring of action of the Deity was Rectitude, and Wisdom.

In 1736 John Noon published a tract entitled 'An Introduction to the Doctrine of Fluxions, and a Defense of the Mathematicians against the objections of the Author of the Analyst'. De Morgan (1860) says: 'This very acute tract is anonymous, but it was always attributed to Bayes by the contemporaries who write in the names of the authors as I have seen in various copies, and it bears his name in other places.' The ascription to Bayes is accepted also in the British Museum catalogue.

From the copy in Dr. Williams's library we quote:

> [p. 9]: It is not the business of the Mathematician to dispute whether quantities do in fact ever vary in the manner that is supposed, but only whether the notion of their doing so be intelligible; which being allowed, he has a right to take it for granted, and then see what deductions he can make from that supposition. It is not the business of a Mathematician to show that a strait line or circle can be drawn, but he tells you what he means by these; and if you understand him, you may proceed further with him; and it would not be to the purpose to object that there is no such thing in nature as a true strait line or perfect circle, for this is none of his concern: he is not inquiring how things are in matter of fact, but supposing things to be in a certain way, what are the consequences to be deduced from them; and all that is to be demanded of him is, that his suppositions be intelligible, and his inferences just from the suppositions he makes.
>
> [p. 48]: He [i.e. the Analyst = Bishop Berkeley] represents the disputes and controversies among mathematicians as disparaging the evidence of their meth-

> ods: and, Query 51, he represents Logics and Metaphysics as proper to open their eyes, and extricate them from their difficulties. Now were ever two things thus put together? If the disputes of the professors of any science disparage the science itself, Logics and Metaphysics are much more to be disparaged than Mathematics; why, therefore, if I am half blind, must I take for my guide one that can't see at all?
>
> [p. 50]: So far as Mathematics do not tend to make men more sober and rational thinkers, wiser and better men, they are only to be considered as an amusement, which ought not to take us off from serious business.

This tract may have had something to do with Bayes's election, in 1742, to Fellowship of the Royal Society, for which his sponsors were Earl Stanhope, Martin Folkes, James Burrow, Cromwell Mortimer, and John Eames.

William Whiston, Newton's successor in the Lucasian Chair at Cambridge, who was expelled from the University for Arianism, notes in his Memoirs (p. 390) that 'on August the 24th this year 1746, being Lord's Day, and St. Bartholomew's Day, I breakfasted at Mr Bay's, a dissenting Minister at Tunbridge Wells, and a Successor, though not immediate, to Mr Humphrey Ditton, and like him a very good mathematician also'. Whiston goes on to relate what he said to Bayes, but he gives no indication that Bayes made reply.

According to Strange (1949) Bayes wished to retire from his ministry as early as 1749, when he allowed a group of Independents to bring ministers from London to take services in his chapel week by week, except for Easter, 1750, when he refused his pulpit to one of these preachers; and in 1752 he was succeeded in his ministry by the Rev. William Johnston, A.M., who inherited Bayes's valuable library. Bayes continued to live in Tunbridge Wells until his death on 17 April 1761. His body was taken to be buried, with that of his father, mother, brothers and sisters, in the Bayes and Cotton family vault in Bunhill Fields, the Nonconformist burial ground by Moorgate. This cemetery also contains the grave of Bayes's friend, the Unitarian Rev. Richard Price, author of the *Northampton Life Table* and object of Burke's oratory and invective in *Reflections on the French Revolution*, and the graves of John Bunyan, Samuel Watts, Daniel Defoe, and many other famous men.

Bayes's will, executed on 12 December 1760, shows him to have been a man of substance. The bulk of his estate was divided among his brothers, sisters, nephews and cousins, but he left £200 equally between 'John Boyl late preacher at Newington and now at Norwich, and Richard Price now I suppose preacher at Newington Green'. He also left 'To Sarah Jeffrey daughter of John Jeffrey, living with her father at the corner of Fountains Lane near Tunbridge Wells, £500, and my watch made by Elliott and all my linen and wearing apparell and household stuff.'

Apart from the tracts already noted, and the celebrated Essay reproduced here, Bayes wrote a letter on Asymptotic Series to John Canton, published in the *Philosophical Transactions of the Royal Society* (1763, pp. 269–271). His mathematical work, though small in quantity, is of the very highest quality; both his tract on fluxions and his paper on asymptotic series contain thoughts which did not receive as clear expression again until almost a century had elapsed.

Since copies of the volume in which Bayes's essay first appeared are not rare, and copies of a photographic reprint issued by the Department of Agriculture, Washington, D.C., U.S.A., are fairly widely dispersed, the view has been taken that in preparing Bayes's paper for publication here some editing is permissible. In particular, the notation has been modernized, some of the archaisms have been removed and what seem to be obvious printer's errors have been corrected. Sometimes, when a word has been omitted in the original, a suggestion has been supplied, enclosed in square brackets. Otherwise, however, nothing has been changed, and we hope that while the present text should in no sense be regarded as definitive, it will be easier to read on that account. All the work of preparing the text for the printer was most painstakingly and expertly carried out by Mr M. Gilbert, B.Sc., A.R.C.S. Thanks are also due to the Royal Society for permission to reproduce the Essay in its present form.

In writing the biographical notes the present author has had the friendly help of many persons, including especially Dr A. Fletcher and Mr R. L. Plackett, of the University of Liverpool, Mr J. F. C. Willder, of the Department of Pathology, Guy's Hospital Medical School, and Mr M. E. Ogborn, F.I.A., of the Equitable Life Assurance Society. He would also like to thank Sir Ronald Fisher, for some initial prodding which set him moving, and Prof. E. S. Pearson, for patient encouragement to see the matter through to completion.

REFERENCES

ANDERSON J. G. (1941). *Mathematical Gazette*, **25**, 160–2.

CANTOR, M. (1908). *Geschichte der Mathematik*, vol. IV. (Article by Netto.)

DE MORGAN, A. (1860). *Notes and Queries*, 7 Jan. 1860.

LORIA, G. (1933), *Storia delle Mathematiche*, vol. III. Turin.

MACKENZIE, M. (Ed.) (1865). *Imperial Dictionary of Universal Biography*, 3 vols., Glasgow.

STRANGE, C. H. (1949). *Nonconformity in Tunbridge Wells*. Tunbridge Wells.

The Gentleman's Magazine (1761), **31**, 188.

Notes and Queries (1941), 19 April.

APPENDIX 3

An Essay Towards Solving a Problem in the Doctrine of Chances

by The Late Rev. Mr Bayes, F.R.S.

Communicated by Mr Price, in a Letter to John Canton, A.M., F.R.S.

Read 23 December 1763

Dear Sir,

I now send you an essay which I have found among the papers of our deceased friend Mr Bayes, and which, in my opinion, has great merit, and well deserves to be preserved. Experimental philosophy, you will find, is nearly interested in the subject of it; and on this account there seems to be particular reason for thinking that a communication of it to the Royal Society cannot be improper.

He had, you know, the honour of being a member of that illustrious Society, and was much esteemed by many in it as a very able mathematician. In an introduction which he has writ to this Essay, he says, that his design at first in thinking on the subject of it was, to find out a method by which we might judge concerning the probability that an event has to happen, in given circumstances, upon supposition that we know nothing concerning it but that, under the same circumstances, it has happened a certain number of times, and failed a certain other number of times. He adds, that he soon perceived that it would not be very difficult to do this, provided some rule could be found according to which we ought to estimate the chance that the probability for the happening of an event perfectly unknown, should lie between any two named degrees of probability, antecedently to any experiments made about it; and that it appeared to him that the rule must be to suppose the chance the same that it should lie between any two equidifferent degrees; which, if it were allowed, all the rest might be easily calculated in the common method of proceeding in the doctrine of chances. Accordingly, I find among his papers a very ingenious solution of this problem in this way. But he afterwards considered, that the *postulate* on which he had argued might not perhaps be looked upon by all as reasonable; and therefore he chose to lay down in another form the proposition in which he thought the solution of the problem is contained, and in a *scholium* to subjoin the reasons why he thought so, rather than to take into his mathematical reasoning any thing that might admit dispute. This, you will observe, is the method which he has pursued in this essay.

Every judicious person will be sensible that the problem now mentioned is by no means merely a curious speculation in the doctrine of chances, but necessary to be solved in order to [provide] a sure foundation for all our reasonings concerning past facts, and what is likely to be hereafter. Common sense is indeed sufficient to shew us that, from the observation of what has in former instances been the consequence of a certain cause or action, one may make a judgment what is likely to be the consequence of it another time, and that the larger [the] number of experiments we have to support a conclusion, so much the more reason we have to take it for granted. But it is certain that we cannot determine, at least not to any nicety, in what degree repeated experiments confirm a conclusion, without the particular discussion of the beforementioned problem; which, therefore, is necessary to be considered by any one who would give a clear account of the strength of *analogical* or *inductive reasoning*; concerning, which at present, we seem to know little more than that it does sometimes in fact convince us, and at other times not; and that, as it is the means of [a]cquainting us with many truths, of which otherwise we must have been ignorant; so it is, in all probability, the source of many errors, which perhaps might in some measure be avoided, if the force that this sort of reasoning ought to have with us were more distinctly and clearly understood.

These observations prove that the problem enquired after in this essay is no less important than it is curious. It may be safely added, I fancy, that it is also a problem that has never before been solved. Mr De Moivre, indeed, the great improver of this part of mathematics, has in his *Laws of Chance*,* after Bernoulli, and to a greater degree of exactness, given rules to find the probability there is, that if a very great number of trials be made concern-

*See Mr De Moivre's *Doctrine of Chances*, p. 243, etc. He has omitted the demonstrations of his rules, but these have been since supplied by Mr Simpson at the conclusion of his treatise on *The Nature and Laws of Chance*.

ing any event, the proportion of the number of times it will happen, to the number of times it will fail in those trails, should differ less than by small assigned limits from the proportion of the probability of its happening to the probability of its failing in one single trial. But I know of no person who has shewn how to deduce the solution of the converse problem to this; namely, 'the number of times an unknown event has happened and failed being given, to find the chance that the probability of its happening should lie somewhere between any two named degrees of probability.' What Mr De Moivre has done therefore cannot be thought sufficient to make the consideration of this point unnecessary: especially, as the rules he has given are not pretended to be rigorously exact, except on supposition that the number of trials made are infinite; from whence it is not obvious how large the number of trials must be in order to make them exact enough to be depended on in practice.

Mr De Moivre calls the problem he has thus solved, the hardest that can be proposed on the subject of chance. His solution he has applied to a very important purpose, and thereby shewn that those are much mistaken who have insinuated that the Doctrine of Chances in mathematics is of trivial consequence, and cannot have a place in any serious enquiry.† The purpose I mean is, to shew what reason we have for believing that there are in the constitution of things fixt laws according to which events happen, and that, therefore, the frame of the world must be the effect of the wisdom and power of an intelligent cause; and thus to confirm the argument taken from final causes for the existence of the Deity. It will be easy to see that the converse problem solved in this essay is more directly applicable to this purpose; for it shews us, with distinctness and precision, in every case of any particular order or recurrency of events, what reason there is to think that such recurrency or order is derived from stable causes or regulations in nature, and not from any of the irregularities of chance.

The two last rules in this essay are given without the deductions of them. I have chosen to do this because these deductions, taking up a good deal of room, would swell the essay too much; and also because these rules, though of considerable use, do not answer the purpose for which they are given as perfectly as could be wished. They are, however, ready to be produced, if a communication of them should be thought proper. I have in some places writ short notes, and to the whole I have added an application of the rules in the essay to some particular cases, in order to convey a clearer idea of the nature of the problem, and to shew how far the solution of it has been carried.

†See his *Doctrine of Chances*, p. 252, etc.

I am sensible that your time is so much taken up that I cannot reasonably expect that you should minutely examine every part of what I now send you. Some of the calculations, particularly in the Appendix, no one can make without a good deal of labour. I have taken so much care about them, that I believe there can be no material error in any of them; but should there be any such errors, I am the only person who ought to be considered as answerable for them.

Mr Bayes has thought fit to begin his work with a brief demonstration of the general laws of chance. His reason for doing this, as he says in his introduction, was not merely that his reader might not have the trouble of searching elsewhere for the principles on which he has argued, but because he did not know whither to refer him for a clear demonstration of them. He has also made an apology for the peculiar definition he has given of the word *chance* or *probability*. His design herein was to cut off all dispute about the meaning of the word, which in common language is used in different senses by persons of different opinions, and according as it is applied to *past* or *future* facts. But whatever different senses it may have, all (he observes) will allow that an expectation depending on the truth of any *past* fact, or the happening of any *future* event, ought to be estimated so much the more valuable as the fact is more likely to be true, or the event more likely to happen. Instead therefore, of the proper sense of the word *probability*, he has given that which all will allow to be its proper measure in every case where the word is used. But it is time to conclude this letter. Experimental philosophy is indebted to you for several discoveries and improvements; and, therefore, I cannot help thinking that there is a peculiar propriety in directing to you the following essay and appendix. That your enquiries may be rewarded with many further successes, and that you may enjoy every valuable blessing, is the sincere wish of, Sir,

your very humble servant,

RICHARD PRICE

Newington-Green,
10 *November* 1763

APPENDIX 4

An Essay Towards Solving a Problem in the Doctrine of Chances

by Reverend Thomas Bayes

PROBLEM

Given the number of times in which an unknown event has happened and failed: *Required* the chance that the probability of its happening in a single trial lies somewhere between any two degrees of probability that can be named.

SECTION I

Definition

1. Several events are *inconsistent*, when if one of them happens, none of the rest can.
2. Two events are *contrary* when one, or other of them must; and both together cannot happen.
3. An event is said to *fail*, when it cannot happen; or, which comes to the same thing, when its contrary has happened.
4. An event is said to be determined when it has either happened or failed.
5. The *probability of any event* is the ratio between the value at which an expectation depending on the happening of the event ought to be

computed, and the value of the thing expected upon it's* happening.

6. By *chance* I mean the same as probability.
7. Events are independent when the happening of any one of them does neither increase nor abate the probability of the rest.

Proposition 1

When several events are inconsistent the probability of the happening of one or other of them is the sum of the probabilities of each of them.

Suppose there be three such events, and whichever of them happens I am to receive N, and that the probability of the 1st, 2nd, and 3rd are respectively a/N, b/N, c/N. Then (by the definition of probability) the value of my expectation from the 1st will be a, from the 2nd b, and from the 3rd c. Wherefore the value of my expectations from all three will be $a + b + c$. But the sum of my expectations from all three is in this case an expectation of receiving N upon the happening of one or other of them. Wherefore (by definition 5) the probability of one or other of them is $(a + b + c)/N$ or $a/N + b/N + c/N$, the sum of the probabilities of each of them.

Corollary

If it be certain that one or other of the three events must happen, then $a + b + c = N$. For in this case all the expectations together amounting to a certain expectation of receiving N, their values together must be equal to N. And from hence it is plain that the probability of an event added to the probability of its failure (or of its contrary) is the ratio of equality. For these are two inconsistent events, one of which necessarily happens. Wherefore if the probability of an event is P/N that of it's failure will be $(N - P)/N$.

Proposition 2

If a person has an expectation depending on the happening of an event, the probability of the event is to the probability of its failure as his loss if it fails to his gain if it happens.

Suppose a person has an expectation of receiving N, depending on an event the probability of which is P/N. Then (by definition 5) the value of his expectation is P, and therefore if the event fail, he loses that which in value is P; and if it happens he receives N, but his expectation ceases. His

*Author's Note: the spelling "it's" was correct and appropriate form in Bayes' time even though today we would use "its."

gain therefore is $N - P$. Likewise since the probability of the event is P/N, that of its failure (by corollary prop. 1) is $(N - P)/N$. But P/N is to $(N - P)/N$ as P is to $N - P$, i.e. the probability of the event is to the probability of it's failure, as his loss if it fails to his gain if it happens.

Proposition 3

The probability that two subsequent events will both happen is a ratio compounded of the probability of the 1st, and the probability of the 2nd on supposition the 1st happens.

Suppose that, if both events happen, I am to receive N, that the probability both will happen is P/N, that the 1st will is a/N (and consequently that the 1st will not is $(N - a)/N$) and that the 2nd will happen upon supposition the 1st does is b/N. Then (by definition 5) P will be the value of my expectation, which will become b if the 1st happens. Consequently if the 1st happens, my gain by it is $b - P$, and if it fails my loss is P. Wherefore, by the foregoing proposition, a/N is to $(N - a)/N$, i.e. a is to $N - a$ as P is to $b - P$. Wherefore (*componendo inverse*) a is to N as P is to b. But the ratio of P to N is compounded of the ratio of P to b, and that of b to N. Wherefore the same ratio of P to N is compounded of the ratio of a to N and that of b to N, i.e. the probability that the two subsequent events will both happen is compounded of the probability of the 1st and the probability of the 2nd on supposition the 1st happens.

Corollary

Hence if of two subsequent events the probability of the 1st be a/N, and the probability of both together be P/N, then the probability of the 2nd on supposition the 1st happens is P/a.

Proposition 4

If there be two subsequent events to be determined every day, and each day the probability of the 2nd is b/N and the probability of both P/N, and I am to receive N if both the events happen the first day on which the 2nd does; I say, according to these conditions, the probability of my obtaining N is P/b. For if not, let the probability of my obtaining N be x/N and let y be to x as $N - b$ to N. Then since x/N is the probability of my obtaining N (by definition 1) x is the value of my expectation. And again, because according to the foregoing conditions the first day I have an expectation of obtaining N depending on the happening of both the events together, the probability of which is P/N, the value of this expectation is

P. Likewise, if this coincident should not happen I have an expectation of being reinstated in my former circumstances, i.e. of receiving that which in value is $\dot{x}$ depending on the failure of the 2nd event the probability of which (by corollary to proposition 1) is $(N - b)/N$ or y/x, because y is to x as $N - b$ to N. Wherefore since x is the thing expected and y/x the probability of obtaining it, the value of this expectation is y. But these two last expectations together are evidently the same with my original expectation, the value of which is x, and therefore $P + y = x$. But y is to x as $N - b$ is to N. Wherefore x is to P as N is to b, and x/N (the probability of my obtaining N) is P/b.

Corollary

Suppose after the expectation given me in the foregoing proposition, and before it is at all known whether the 1st event has happened or not, I should find that the 2nd event has happened; from hence I can only infer that the event is determined on which my expectation depended, and have no reason to esteem the value of my expectation either greater or less than it was before. For if I have reason to think it less, it would be reasonable for me to give something to be reinstated in my former circumstances, and this over and over again as often as I should be informed that the 2nd event had happened, which is evidently absurd. And the like absurdity plainly follows if you say I ought to set a greater value on my expectation than before, for then it would be reasonable for me to refuse something if offered me upon condition I would relinquish it, and be reinstated in my former circumstances; and this likewise over and over again as often as (nothing being known concerning the 1st event) it should appear that the 2nd had happened. Notwithstanding therefore this discovery that the 2nd event has happened, my expectation ought to be esteemed the same in value as before, i.e. x, and consequently the probability of my obtaining N is (by definition 5) still x/N or P/b.* But after this discovery the probability of my obtaining N is the probability that the 1st of two subsequent events has happened upon the supposition that the 2nd has, whose probabilities were as before specified. But the probability that an event has happened is the same as the probability I have to guess right if I guess it has happened. Wherefore the following proposition is evident.

*What is here said may perhaps be a little illustrated by considering that all that can be lost by the happening of the 2nd event is the chance I should have had of being reinstated in my former circumstances, if the event on which my expectation depended had been determined in the manner expressed in the proposition. But this chance is always as much *against* me as it is *for* me. If the 1st event happens, it is *against* me, and equal to the chance for the 2nd event's failing. If the 1st event does not happen, it is *for* me, and equal also to the chance for the 2nd event's failing. The loss of it, therefore, can be no disadvantage.

Proposition 5

If there be two subsequent events, the probability of the 2nd b/N and the probability of both together P/N, and it being first discovered that the 2nd event has happened, from hence I guess that the 1st event has also happened, the probability I am in the right is P/b.*

Proposition 6

The probability that several independent events shall all happen is a ratio compounded of the probabilities of each.

For from the nature of independent events, the probability that any one happens is not altered by the happening or failing of any of the rest, and consequently the probability that the 2nd event happens on supposition the 1st does is the same with its original probability; but the probability that any two events happen is a ratio compounded of the probability of the 1st event, and the probability of the 2nd on supposition the 1st happens by proposition 3. Wherefore the probability that any two independent events both happen is a ratio compounded of the probability of the 1st and the probability of the 2nd. And in like manner considering the 1st and 2nd events together as one event; the probability that three independent events all happen is a ratio compounded of the probability that the two 1st both happen and the probability of the 3rd. And thus you may proceed if there be ever so many such events; from whence the proposition is manifest.

Corollary 1

If there be several independent events, the probability that the 1st happens the 2nd fails, the 3rd fails and the 4th happens, etc. is a ratio compounded of the probability of the 1st, and the probability of the failure of the 2nd, and the probability of the failure of the 3rd, and the probability of the 4th, etc. For the failure of an event may always be considered as the happening of its contrary.

*What is proved by Mr Bayes in this and the preceding proposition is the same with the answer to the following question. What is the probability that a certain event, when it happens, will be accompanied with another to be determined at the same time? In this case, as one of the events is given, nothing can be due for the expectation of it; and, consequently, the value of an expectation depending on the happening of both events must be the same with the value of an expectation depending on the happening of one of them. In other words; the probability that, when one of two events happens, the other will, is the same with the probability of this other. Call x then the probability of this other, and if b/N be the probability of the given event, and p/N the probability of both, because $p/N = (b/N) \times x$, $x = p/b =$ the probability mentioned in these propositions.

Corollary 2

If there be several independent events, and the probability of each one be a, and that of its failing be b, the probability that the 1st happens and the 2nd fails, and the 3rd fails and the 4th happens, etc. will be *abba*, etc. For, according to the algebraic way of notation, if a denote any ratio and b another, *abba* denotes the ratio compounded of the ratios a, b, b, a. This corollary therefore is only a particular case of the foregoing.

Definition

If in consequence of certain data there arises a probability that a certain event should happen, its happening or failing, in consequence of these data, I call it's happening or failing in the 1st trial. And if the same data be again repeated, the happening or failing of the event in consequence of them I call its happening or failing in the 2nd trail; and so on as often as the same data are repeated. And hence it is manifest that the happening or failing of the same event in so many diffe[rent] trials, is in reality the happening or failing of so many distinct independent events exactly similar to each other.

Proposition 7

If the probability of an event be a, and that of its failure be b in each single trial, the probability of its happening p times, and failing q times in $p + q$ trails is Ea^pb^q if E be the coefficient of the term in which occurs a^pb^q when the binomial $(a + b)^{p+q}$ is expanded.

For the happening or failing of an event in different trials are so many independent events. Wherefore (by cor. 2 prop. 6) the probability that the event happens the 1st trial, fails the 2nd and 3rd, and happens the 4th, fails the 5th, etc. (thus happening and failing till the number of times it happens be p and the number if fails be q) is *abbab* etc. till the number of a's be p and the number of b's be q, that is; 'tis a^pb^q. In like manner if you consider the event as happening p times and failing q times in any other particular order, the probability for it is a^pb^q; but the number of different orders according to which an event may happen or fail, so as in all to happen p times and fail q, in $p + q$ trails is equal to the number of permutations that *aaaa bbb* admit of when the number of a's is p, and the number of b's is q. And this number is equal to E, the coefficient of the term in which occurs a^pb^q when $(a + b)^{p+q}$ is expanded. The event therefore may happen p times and fail q in $p + q$ trails E different ways and no more, and its happening and failing these several different ways are so many inconsistent events, the probability for each of which is a^pb^q, and therefore by prop. 1 the probability that some way or other it happens p times and fails q times in $p + q$ trials is $Ea^p b^q$.

SECTION II

Postulate

1. I suppose the square table or plane *ABCD* to be so made and levelled, that if either of the balls *o* or *W* be thrown upon it, there shall be the same probability that it rests upon any one equal part of the plane as another, and that it must necessarily rest somewhere upon it.

2. I suppose that the ball *W* shall be first thrown, and through the point where it rests a line *os* shall be drawn parallel to *AD*, and meeting *CD* and *AB* in *s* and *o*; and that afterwards the ball *O* shall be thrown $p + q$ or n times, and that its resting between *AD* and *os* after a single throw be called the happening of the event *M* in a single trial. These things supposed:

Lemma 1

The probability that the point *o* will fall between any two points in the line *AB* is the ratio of the distance between the two points to the whole line *AB*.

Let any two points be named, as *f* and *b* in the line *AB*, and through them parallel to *AD* draw *fF*, *bL* meeting *CD* in *F* and *L*. Then if the rectangles *Cf*, *Fb*, *LA* are commensurable to each other, they may each be divided into the same equal parts, which being done, and the ball *W* thrown, the probability it will rest somewhere upon any number of these equal parts will be the sum of the probabilities it has to rest upon each one of them, because its resting upon any different parts of the plane *AC* are so many inconsistent events; and this sum, because the probability it should rest upon any one equal part as another is the same, is the probability it should rest upon any one equal part multiplied by the number of parts. Consequently, the probability there is that the ball *W* should rest somewhere upon *Fb* is the probability it has to rest upon one equal part multiplied by the number of equal parts in *Fb*; and the probability it rests somewhere upon *Cf* or *LA*, i.e. that it does not rest upon *FB* (because it must rest somewhere upon *AC*) is the probability it rests upon one equal part multiplied by the number of equal parts in *Cf*, *LA* taken together. Wherefore, the probability it rests upon *Fb* is to the probability it does not as the number of equal parts in *Fb* is to the number of equal parts in *Cf*, *LA* together, or as *Fb* to *Cf*, *LA* together, or as *fb* to *Bf*, *Ab* together. Wherefore the probability it rests upon *Fb* is to the probability it does not as *fb* to *Bf*, *Ab* together. And (*componendo inverse*) the probability it rests upon *Fb* is to the probability it rests upon *Fb* added to the probability it does not, as *fb* to *AB*, or as the ratio of *fb* to *AB* to the ratio of *AB* to *AB*. But the probability of any event added to the probability of its failure is the ratio of equality; wherefore, the probability it rests upon *Fb* is to the ratio of equality as the ratio of *fb* to *AB* to the ratio of *AB* to *AB*, or the ratio of equality; and therefore the probability it rests upon *Fb* is the ratio of *fb* to *AB*. But *ex hypothesi* according as the ball *W* falls upon *Fb* or not the point *o* will lie between *f* and *b* or not, and therefore the probability the point *o* will lie between *f* and *b* is the ratio of *fb* to *AB*.

Again; if the rectangles *Cf*, *Fb*, *LA* are not commensurable, yet the last mentioned probability can be neither greater nor less than the ratio of *fb* to *AB*; for, if it be less, let it be the ratio of *fc* to *AB*, and upon the line *fb* take the points *p* and *t*, so that *pt* shall be greater than *fc*, and the three lines *Bp*, *pt*, *tA* commensurable (which it is evident may be always done by dividing *AB* into equal parts less than half *cb*, and taking *p* and *t* the nearest points of division to *f* and *c* that lie upon *fb*). Then because *Bp*, *pt*, *tA* are commensurable, so are the rectangles *Cp*, *Dt*, and that upon *pt* compleating the square *AB*. Wherefore, by what has been said, the probability that the point *o* will lie between *p* and *t* is the ratio of *pt* to *AB*. But if it lies between *p* and *t* it must lie between *f* and *b*. Wherefore, the probability it should lie between *f* and *b* cannot beless than the ratio of *pt* to *AB*, and therefore must be greater than the ratio of *fc* to *AB* (since *pt* is greater than *fc*). And after the same manner you may prove that the forementioned probability cannot be greater than the ratio of *fb* to *AB*, it must therefore be the same.

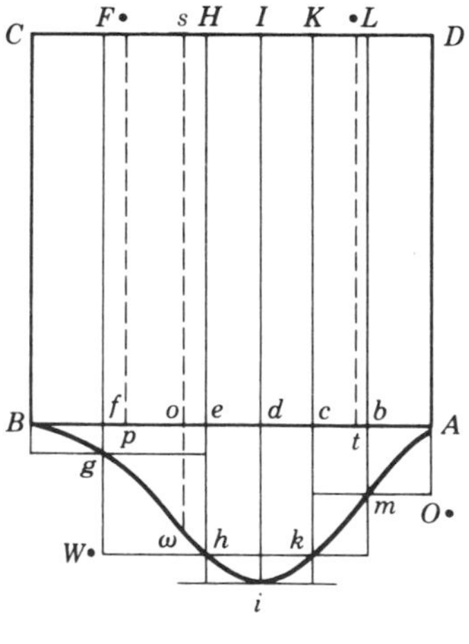

Lemma 2

The ball W having been thrown, and the line os drawn, the probability of the event M in a single trial is the ratio of Ao to AB.

For, in the same manner as in the foregoing lemma, the probability that the ball o being thrown shall rest somewhere upon Do or between AD and so is the ratio of Ao to AB. But the resting of the ball o between AD and so after a single throw is the happening of the event M in a single trials. Wherefore the lemma is manifest.

Proposition 8

If upon BA you erect the figure $BghikmA$ whose property is this, that (the base BA being divided into any two parts, as Ab, and Bb and at the point of division b a perpendicular being erected and terminated by the figure in m; and y, x, r representing respectively the ratio of bm, Ab, and Bb to AB, and E being the coefficient of the term in which occurs $a^p b^q$ when the binomial $(a + b)^{p+q}$ is expanded) $y = Ex^p r^q$. I say that before the ball W is thrown, the probability the point o should fall between f and b, any two points named in the line AB, and with all that the event M should happen p times and fail q in $p + q$ trials, is the ratio of $fghikmb$, the part of the figure $BghikmA$ intercepted between the perpendiculars fg, bm raised upon the line AB, to CA the square upon AB.

DEMONSTRATION

For if not; first let it be the ratio of D a figure greater than $fghikmb$ to CA, and through the points e, d, c draw perpendiculars of fb meeting the curve $AmigB$ in h, i, k; the point d being so placed that di shall be the longest of the perpendiculars terminated by the line fb, and the curve $AmigB$; and the points e, d, c being so many and so placed that the rectangles, bk, ci, ei, fh taken together shall differ less from $fghikmb$ than D does; all which may be easily done by the help of the equation of the curve, and the difference between D and the figure $fghikmb$ given. Then since di is the longest of the perpendicular ordinates that insist upon fb, the rest will gradually decrease as they are farther and farther from it on each side, as appears from the construction of the figure, and consequently eh is greater than gf or any other ordinate that insists upon ef.

Now if Ao were equal to Ae, then by lem. 2 the probability of the event M in a single trial would be the ratio of Ae to AB, and consequently by cor. Prop. 1 the probability of it's failure would be the ratio of Be to AB. Wherefore, if x and r be the two forementioned ratios respectively, by proposition 7 the probability of the event M happening p times and failing q in $p + q$ trials would be $Ex^p r^q$. But x and r being respectively the ratios of Ae to AB and Be to AB, if y is the ratio of eh to AB, then, by construction of the figure AiB, $y = Ex^p r^q$. Wherefore, if Ao were equal to Ae the probability of the event M happening p times and failing q in $p + q$ trials would be y, or the ratio of eh to AB. And if Ao were equal to Af, or were any mean between Ae and Af, the last mentioned probability for the same reasons would be the ratio of fg or some other of the ordinates insisting upon ef, to AB. But eh is the greatest of all the ordinates that insist upon ef. Wherefore, upon supposition the point should lie anywhere between f and e, the probability that the event M happens p times and fails q in $p + q$ trials cannot be greater than the ratio of eh to AB. There then being these two subsequent events, the 1st that the point o will lie between e and f, the 2nd that the event M will happen p times and fail q in $p + q$ trails, and the probability of the first (by lemma 1) is the ratio of ef to AB, and upon supposition the 1st happens, by what has been now proved, the probability of the 2nd cannot be greater than the ratio of eh to AB, it evidently follows (from proposition 3) that the probability both together will happen cannot be greater than the ratio compounded of that of ef to AB and that of eh to AB, which compound ratio is the ratio of fh to CA. Wherefore, the probability that the point o will lie between f and e, and the event M happen p times and fail q, is not greater than the ratio of fh to CA. And in like manner the probability the point o will lie between e and d, and the event M happen and fails as before, cannot be greater than the ratio of ei to CA. And again, the probability the point o will lie between d and c, and the event M happen and fail as before, cannot be greater than the ratio of ci to CA. And lastly, the probability that the point o will lie between c and b, and the event M happen and fail as before, cannot be greater than the ratio of bk to CA. Add now all these several probabilities together, and their sum (by proposition 1) will be the probability that the point will lie somewhere between f and b, and the event M happen p times and fail q in $p + q$ trials. Add likewise the correspondent ratios together, and their sum will be the ratio of the sum of the antecedents to their common consequent, i.e. the ratio of fh, ei, ci, bk together to CA; which ratio is less than that of D to CA, because D is greater than fh, ei, ci, bk together. And therefore, the probability that the point o will lie between f and b, and withal [in addition (author's note)] that the event M will happen p times and fail q in $p + q$ trials, is less than the ratio of D to CA; but it was supposed the same which is absurd. And in like manner, by inscribing rectangles within the figure, as eg, dh, dh, dk, cm, you may prove that the last mentioned probability is *greater* than the ratio of any figure less than $fghikmb$ to CA.

Wherefore, that probability must be the ratio of $fghikmb$ to CA.

Corollary

Before the ball W is thrown the probability that the point o will lie somewhere between A and B, or somewhere upon the line AB, and withal that the event M will happen p times, and fail q in $p + q$ trials is the ratio of the whole figure AiB to CA. But it is certain that the point o will lie somewhere upon AB. Wherefore, before the ball W is thrown the probability the event M will happen p times and fail q in $p + q$ trials is the ratio of AiB to CA.

Proposition 9

If before anything is discovered concerning the place of the point o, it should appear that the event M had happened p times and failed q in $p + q$ trials, and from hence I guess that the point o lies between any two points in the line AB, as f and b, and consequently that the probability of the event M in a single trial was somewhere between the ratio of Ab to AB and that of Af to AB; the probability I am in the right is the ratio of that part of the figure AiB described as before which is intercepted between perpendiculars erected upon AB at the points f and b, to the whole figure AiB.

For, there being these two subsequent events, the first that the point o will lie between f and b; the second that the event M should happen p times and fail q in $p + q$ trials; and (by corollary to proposition 8) the original probability of the second is the ratio of AiB to CA, and (by proposition 8) the probability of both is the ratio of $fghimb$ to CA; wherefore (by proposition 5) it being first discovered that the second has happened, and from hence I guess that the first has happened also, the probability I am in the right is the ratio of $fghimb$ to AiB, the point which was to be proved.

Corollary

The same things supposed, if I guess that the probability of the event M lies somewhere between 0 and the ratio of Ab to AB, my chance to be in the right is the ratio of Abm to AiB.

Scholium

From the preceding proposition it is plain, that in the case of such an event as I there call M, from the number of times it happens and fails in a certain number of trials, without knowing anything more concerning it, one may give a guess whereabouts it's probability is, and, by the usual methods computing the magnitudes of the areas there mentioned, see the chance that the guess is right. And that the same rule is the proper one to be used in the case of an event concerning the probability of which we absolutely know nothing antecedently to any trials made concerning it, seems to appear from the following consideration; viz. that concerning such an event I have no reason to think that, in a certain number of trials, it should rather happen any one possible number of times than another. For, on this account, I may justly reason concerning it as if its probability had been at first unfixed, and then determined in such a manner as to give me no reason to think that, in a certain number of trials, it should rather happen any one possible number of times than another. But this is exactly the case of the event M. For before the ball W is thrown, which determines it's probability in a single trial (by corollary to proposition 8), the probability it has to happen p times and fail q in $p + q$ or n trials is the ratio of AiB to CA, which ratio is the same when $p + q$ or n is given, whatever number p is; as will appear by computing the magnitude of AiB by the method of fluxions.* And consequently before the place of the point o is discovered or the number of times the event M has happened in n trials, I can have no reason to think it should rather happen one possible number of times than another.

In what follows therefore I shall take for granted that the rule given concerning the event M in proposition 9 is also the rule to be used in relation to any event concerning the probability of which nothing at all is known antecedently to any trials made or observed concerning it. And such an event I shall call an unknown event. [Author's Note: The "method of fluxions" referred to in this Scholium is, of course, the Newtonian Calculus. The implication, in modern terms, of Proposition 9 and the Scholium, is given in Eq. (1), Appendix 1.]

Corollary

Hence, by supposing the ordinates in the figure AiB to be contracted in the ratio of E to one, which makes no alteration in the proportion of the parts of the figure intercepted between them, and applying what is said of the event M to an unknown event, we have the following proposition, which gives the rules for finding the probability of an event from the number of times it actually happens and fails.

*It will be proved presently in art. 4 by computing in the method here mentioned that AiB contracted in the ratio of E to 1 is to CA as 1 to $(n + 1)E$: from whence it plainly follows that, antecedently to this contraction, AiB must be to CA in the ratio of 1 to $n + 1$, which is a constant ratio when n is given, whatever p is.

Proposition 10

If a figure be described upon any base *AH* (Vid. Fig.) having for its equation $y = x^p r^q$; where y, x, r are respectively the ratios of an ordinate of the figure insisting on the base at right angles, of the segment of the base intercepted between the ordinate and *A* the beginning of the base, and of the other segment of the base lying between the ordinate and the point *H*, to the base as their common consequent. I say then that if an unknown event has happened p times and failed q in $p + q$ trials, and in the base *AH* taking any two points as f and t you erect the ordinates *fC*, *tF* at right angles with it, the chance that the probability of the event lies somewhere between the ratio of *Af* to *AH* and that of *At* to *AH*, is the ratio of *tFCf*, that part of the before-described figure which is intercepted between the two ordinates, to *ACFH* the whole figure insisting on the base *AH*.

This is evident from proposition 9 and the remarks made in the foregoing scholium and corollary.

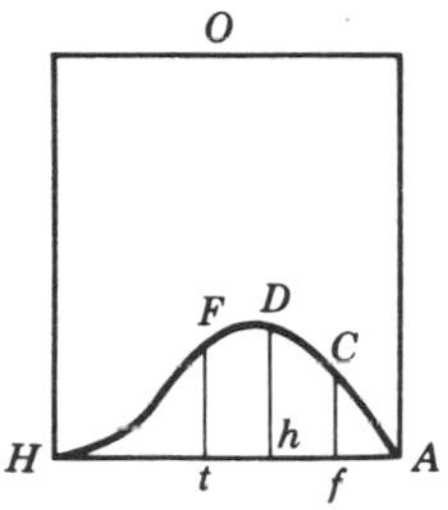

1. Now, in order to reduce the foregoing rule to practice, we must find the value of the area of the figure described and the several parts of it separated, by ordinates perpendicular to its base. For which purpose, suppose $AH = 1$ and *HO* the square upon *AH* likewise $= 1$, and *Cf* will be $= y$, and $Af = x$, and $Hf = r$, because y, x and r denote the ratios of *Cf*, *Af*, and *Hf* respectively to *AH*. And by the equation of the curve $y = x^p r^q$ and (because $Af + fH = AH$) $r + x = 1$. Wherefore

$$y = x^p(1-x)^q$$

$$= x^p - qx^{p+1} + \frac{q(q-1)x^{p+2}}{2} - \frac{q(q-1)(q-2)x^{p+3}}{2.3} + \text{etc.}$$

Now the abscisse being x and the ordinate x^p the correspondent area is

$x^{p+1}/(p+1)$ (by proposition 10, cas. 1, Quadrat. Newt.)† and the ordinate being qx^{p+1} the area is $qx^{p+2}/(p+2)$; and in like manner of the rest. Wherefore, the abscisse being x and the ordinate y or $x^p - qx^{p+1}$ + etc. the correspondent area is

$$\frac{x^{p+1}}{p+1} - \frac{qx^{p+2}}{p+2} + \frac{q(q-1)x^{p+3}}{2(p+3)} - \frac{q(q-1)(q-2)x^{p+4}}{2.3(p+4)} + \text{etc.}$$

[Author's Note: 2.3 means 2 × 3.]
Wherefore, if $x = Af = Af/(AH)$, and $y = Cf = Cf/(AH)$, then

$$ACf = \frac{ACf}{HO} = \frac{x^{p+1}}{p+1} - \frac{qx^{p+2}}{p+2} + \frac{q(q-1)x^{p+3}}{2(p+3)} - \text{etc.}$$

From which equation, if q be a small number, it is easy to find the value of the ratio of ACf to HO and in like manner as that was found out, it will appear that the ratio of HCf to HO is

$$\frac{r^{q+1}}{q+1} - \frac{pr^{q+2}}{q+2} + \frac{p(p-1)r^{q+3}}{2(q+3)} - \frac{p(p-1)(p-2)r^{q+4}}{2.3(q+4)} + \text{etc.}$$

which series will consist of few terms and therefore is to be used when p is small.

2. The same things supposed as before, the ratio of ACf to HO is

$$\frac{x^{p+1}r^q}{p+1} + \frac{qx^{p+2}r^{q-1}}{(p+1)(p+2)} + \frac{q(q-1)x^{p+3}r^{q-2}}{(p+1)(p+2)(p+3)}$$
$$+ \frac{q(q-1)(q-2)x^{p+4}r^{q-3}}{(p+1)(p+2)(p+3)(p+4)}$$
$$+\text{etc.} + \frac{x^{n+1}q(q-1)\cdots 1}{(n+1)(p+1)(p+2)\cdots n},$$

† 'Tis very evident here, without having recourse to Sir Isaac Newton, that the fluxion of the area ACf being

$$y\dot{x} = x^p\dot{x} - qx^{p+1}\dot{x} + \frac{q(q-1)}{2}x^{p+2}\dot{x} - \text{etc.}$$

the fluent or area itself is

$$\frac{x^{p+1}}{p+1} - \frac{qx^{p+2}}{p+2} + \frac{q(q-1)x^{p+3}}{2(p+3)} - \text{etc.}$$

where $n = p + q$. For this series is the same with $x^{p+1}/(p+1) - qx^{p+2}/(p+2) +$ etc. set down in Art. 1st as the value of the ratio of ACf to HO; as will easily be seen by putting in the former instead of r its value $1 - x$, and expanding the terms and ordering them according to the powers of x. Or, more readily, by comparing the fluxions of the two series, and in the former instead of $\dot{r}$ substituting $-\dot{x}$.*

3. In like manner, the ratio of HCf to HO is

$$\frac{r^{q+1}x^p}{q+1} + \frac{pr^{q+2}x^{p-1}}{(q+1)(q+2)} + \frac{p(p-1)r^{q+3}x^{p-2}}{(q+1)(q+2)(q+3)} + \text{etc.}$$

4. If E be the coefficient of that term of the binomial $(a+b)^{p+q}$ expanded in which occurs $a^p b^q$, the ratio of the whole figure $ACFH$ to HO is $\{(n+1)E\}^{-1}$, n being $= p + q$. For, when $Af = AH$, $x = 1$, $r = 0$. Wherefore, all the terms of the series set down in Art. 2 as expressing the

*The fluxion of the first series is

$$x^p r^q \dot{x} + \frac{qx^{p+1}r^{q-1}\dot{r}}{p+1} + \frac{qx^{p+1}r^{q-1}\dot{x}}{p+1} + \frac{q(q-1)x^{p+2}r^{q-2}\dot{r}}{(p+1)(p+2)} + \frac{q(q-1)x^{p+2}r^{q-2}\dot{x}}{(p+1)(p+2)}$$

$$+ \frac{q(q-1)(q-2)x^{p+3}r^{q-3}\dot{r}}{(p+1)(p+2)(p+3)} + \text{etc.}$$

or, substituting $-\dot{x}$ for $\dot{r}$,

$$x^p r^q \dot{x} - \frac{qx^{p+1}r^{q-1}\dot{x}}{p+1} + \frac{qx^{p+1}r^{q-1}\dot{x}}{p+1} - \frac{q(q-1)x^{p+2}r^{q-2}\dot{x}}{(p+1)(p+2)}$$

$$+ \frac{q(q-1)x^{p+2}r^{q-2}\dot{x}}{(p+1)(p+2)} - \text{etc.}$$

which, as all the terms after the first destroy one another, is equal to

$$x^p r^q \dot{x} = x^p(1-x)^q \dot{x} = x^p \dot{x}\left[1 - qx + q\frac{(q-1)}{2}x^2 - \text{etc.}\right]$$

$$= x^p x - qx^{p+1}\dot{x} + \frac{q(q-1)x^{p+2}}{2}\dot{x} - \text{etc.}$$

$$= \text{the fluxion of the latter series, or of } \frac{x^{p+1}}{p+1} - \frac{qx^{p+2}}{p+2} + \text{etc.}$$

The two series therefore are the same.

ratio of ACf to HO will vanish except the last, and that becomes

$$\frac{q(q-1)\cdots 1}{(n+1)(p+1)(p+2)\cdots n}.$$

But E being the coefficient of that term in the binomial $(a+b)^n$ expanded in which occurs $a^p b^q$ is equal to

$$\frac{(p+1)(p+2)\cdots n}{q(q-1)\cdots 1}.$$

And, because Af is supposed to become $= AH$, $ACf = ACH$. From whence this article is plain.

5. The ratio of ACf to the whole figure $ACFH$ is (by Art. 1 and 4)

$$(n+1)E\left[\frac{x^{p+1}}{p+1} - \frac{qx^{p+2}}{p+2} + \frac{q(q-1)x^{p+3}}{2(p+3)} - \text{etc.}\right]$$

and if, as x expresses the ratio of Af to AH, X should express the ratio of At to AH; the ratio of AFt to $ACFH$ would be

$$(n+1)E\left[\frac{X^{p+1}}{p+1} - \frac{qX^{p+2}}{p+2} + \frac{q(q-1)X^{p+3}}{2(p+3)} - \text{etc.}\right]$$

and consequently the ratio of $tFCf$ to $ACFH$ is $(n+1)E$ multiplied into the difference between the two series. Compare this with prop. 10 and we shall have the following practical rule.

Rule 1

If nothing is known concerning an event but that it has happened p times and failed q in $p+q$ or n trials, and from hence I guess that the probability of its happening in a single trial lies somewhere between any two degrees of probability as X and x, the chance I am in the right in my guess is $(n+1)E$ multiplied into the difference between the series

$$\frac{X^{p+1}}{p+1}\,\frac{qX^{p+2}}{p+2} + \frac{q(q-1)X^{p+3}}{2(p+3)} - \text{etc.}$$

and the series

$$\frac{x^{p+1}}{p+1} - \frac{qx^{p+2}}{p+2} + \frac{q(q-1)x^{p+3}}{2(p+3)} - \text{etc.}$$

E being the coefficient of $a^p b^q$ when $(a+b)^n$ is expanded.

This is the proper rule to be used when q is a small number; but if q is large and p small, change everywhere in the series here set down p into q and q into p and x into r or $(1-x)$, and X into $R = (1-X)$; which will not make any alteration in the difference between the two series.

Thus far Mr Bayes's essay.

With respect to the rule here given, it is further to be observed, that when both p and q are very large numbers, it will not be possible to apply it to practice on account of the multitude of terms which the series in it will contain. Mr Bayes, therefore, by an investigation which it would be too tedious to give here, has deduced from this rule another, which is as follows.

Rule 2

If nothing is known concerning an event but that it has happened p times and failed q in $p+q$ or n trials, and from hence I guess that the probability of its happening in a single trial lies between $(p/n)+z$ and $(p/n)-z$; if $m^2 = n^3/(pq)$, $a = p/n$, $b = q/n$, E the coefficient of the term in which occurs $a^p b^q$ when $(a+b)^n$ is expanded, and

$$\Sigma = \frac{(n+1)\surd(2pq)}{n\surd n} E a^p b^q$$

multiplied by the series

$$mz - \frac{m^3 z^3}{3} + \frac{(n-2)m^5 z^5}{2n.5} - \frac{(n-2)(n-4)m^7 z^7}{2n.3n.7} + \frac{(n-2)(n-4)(n-6)m^9 z^9}{2n.3n.4n.9} - \text{etc.}$$

my chance to be in the right is greater than

$$\frac{2\Sigma}{1 + 2Ea^p b^q + 2Ea^p b^{q/n}}*$$

*In Mr Bayes's manuscript this chance is made to be greater than $2\Sigma/(1+2Ea^p b^q)$ and less than $2\Sigma/(1-2Ea^p b^q)$. The third term in the two divisors, as I have given them, being omitted. But this being evidently owing to a small oversight in the deduction of this rule, which I have reason to think Mr Bayes had himself discovered, I have ventured to correct his copy, and to give the rule as I am satisfied it ought to be given.

and less than

$$\frac{2\Sigma}{1 - 2Ea^pb^q - 2Ea^pb^q/n},$$

and if $p = q$ my chance is 2Σ exactly. [Author's Note: $2n.2n$ means $(2n) \times (3n)$.]

In order to render this rule fit for use in all cases it is only necessary to know how to find within sufficient nearness the value of Ea^pb^q and also of the series $mz - \frac{1}{3}m^3z^3 +$ etc. With respect to the former Mr Bayes has proved that, supposing K to signify the ratio of the quadrantal arc to its radius, Ea^pb^q will be equal to $\frac{1}{2}\surd n/\surd(Kpq)$ multiplied by the *ratio*, [h], whose *hyberbolic* logarithm is

$$\frac{1}{12}\left[\frac{1}{n} - \frac{1}{p} - \frac{1}{q}\right] - \frac{1}{360}\left[\frac{1}{n^3} - \frac{1}{p^3} - \frac{1}{q^3}\right] + \frac{1}{1260}\left[\frac{1}{n^5} - \frac{1}{p^5} - \frac{1}{q^5}\right]$$
$$- \frac{1}{1680}\left[\frac{1}{n^7} - \frac{1}{p^7} - \frac{1}{q^7}\right] + \frac{1}{1188}\left[\frac{1}{n^9} - \frac{1}{p^9} - \frac{1}{q^9}\right] - \text{etc.}^\dagger$$

where the numeral coefficients may be found in the following manner. Call them A, B, C, D, E etc. Then

$$A = \frac{1}{2.2.3} = \frac{1}{3.4}, \ B = \frac{1}{2.4.5} - \frac{A}{3}, \ C = \frac{1}{2.6.7} - \frac{10B + A}{5},$$
$$D = \frac{1}{2.8.9} - \frac{35C + 21B + A}{7}, \ E = \frac{1}{2.10.11} - \frac{126C + 84D + 36B + A}{9},$$
$$F = \frac{1}{2.12.13} - \frac{462D + 330C + 165E + 55B + A}{11} \text{etc.}$$

†A very few terms of this series will generally give the hyperbolic logarithm to a sufficient degree of exactness. A similar series has been given by Mr De Moivre, Mr Simpson and other eminent mathematicians in an expression for the sum of the logarithms of the numbers 1, 2, 3, 4, 5, to x, which sum they have asserted to be equal to

$$\tfrac{1}{2}\log c + \left(x + \tfrac{1}{2}\right)\log x - x + \frac{1}{12x} - \frac{1}{360x^3} + \frac{1}{1260x^5} - \text{etc.}$$

c denoting the circumference of a circle whose radius is unity. But Mr Bayes, in a preceding paper in this volume, has demonstrated that, though this expression will very nearly approach to the value of this sum when only a proper number of the first terms is taken, the whole series cannot express any quantity at all, because, let x be what it will, there will always be a part of the series where it will begin to diverge. This observation, though it does not much affect the use of this series, seems well worth the notice of mathematicians.

where the coefficients of B, C, D, E, F, etc. in the values of D, E, F, etc. are the 2, 3, 4, etc. highest coefficients in $(a+b)^7$, $(a+b)^9$, $(a+b)^{11}$, etc. expanded; affixing in every particular value the least of these coefficients to B, the next in magnitude to the furthest letter from B, the next to C, the next to the furthest but one, the next to D, the next to the furthest but two, and so on.*

With respect to the value of the series

$$mz - \tfrac{1}{3}m^3z^3 + \frac{(n-2)m^5z^5}{2n.5} \text{etc.}$$

he has observed that it may be calculated directly when mz is less than 1, or even not greater than $\sqrt{3}$: but when mz is much larger it becomes impracticable to do this; in which case he shews a way of easily finding two values of it very nearly equal between which its true value must lie.

The theorem he gives for this purpose is as follows.

Let K, as before, stand for the ratio of the quadrantal arc to its radius, and H for the ratio whose hyperbolic logarithm is

$$\frac{2^2-1}{2n} - \frac{2^4-1}{360n^3} + \frac{2^6-1}{1260n^5} - \frac{2^8-1}{1680n^7} + \text{etc.}$$

Then the series $mz - \frac{1}{3}m^3z^3$ + etc. will be greater or less than the series

$$\frac{Hn\sqrt{K}}{(n+1)\sqrt{2}} - \frac{n\left(1-\dfrac{2m^2z^2}{n}\right)^{\frac{1}{2}n+1}}{(n+2)2mz} + \frac{n^2\left(1-\dfrac{2m^2z^2}{n}\right)^{\frac{1}{2}n+2}}{(n+2)(n+4)4m^3z^3}$$

$$- \frac{3n^3\left(1-\dfrac{2m^2z^2}{n}\right)^{\frac{1}{2}n+3}}{(n+2)(n+4)(n+6)8m^5z^5}$$

$$+ \frac{3.5.n^4\left(1-\dfrac{2m^2z^2}{n}\right)^{\frac{1}{2}n+4}}{(n+2)(n+4)(n+6)(n+8)16m^7z^7} - \text{etc.}$$

*This method of finding these coefficients I have deduced from the demonstration of the third lemma at the end of Mr Simpson's *Treatise on the Nature and Laws of Chance*.

continued to any number of terms, according as the last term has a positive or a negative sign before it.

From substituting these values of Ea^pb^q and

$$mz - \frac{m^3z^3}{3} + \frac{(n-2)}{2n}\frac{m^5z^5}{5}\text{etc.}$$

in the second rule arises a third rule, which is the rule to be used when mz is of some considerable magnitude.

Rule 3

If nothing is known of an event but that it has happened p times and failed q in $p+q$ or n trials, and from hence I judge that the probability of its happening in a single trial lies between $p/n+z$ and $p/n-z$ my chance to be right is *greater* than

$$\frac{\frac{1}{2}\surd(Kpq)h}{\surd(Kpq)+hn^{\frac{1}{2}}+hn^{-\frac{1}{2}}}\left\{2H - \frac{\surd 2(n+1)(1-2m^2z^2/n)^{\frac{1}{2}n+1}}{\surd K(n+2)mz}\right\}$$

and *less* than

$$\frac{\frac{1}{2}\surd(Kpq)h}{\surd(Kpq)-hn^{\frac{1}{2}}-hn^{-\frac{1}{2}}}\left\{2H - \frac{\surd 2(n+1)(1-2m^2z^2/n)^{\frac{1}{2}n+1}}{\surd K(n+2)mz} + \frac{\surd 2n(n+1)(1-2m^2z^2/n)^{\frac{1}{2}n+z}}{\surd K(n+2)(n+4)2m^3z^3}\right\}$$

where m^2, K, h and H stand for the quantities already explained.

AN APPENDIX

Containing an application of the foregoing Rules to some particular Cases

The first rule gives a direct and perfect solution in all cases; and the two following rules are only particular methods of approximating to the solution given in the first rule, when the labour of applying it becomes too great.

The first rule may be used in all cases where either p or q are nothing or not large. The second rule may be used in all cases where mz is less than $\surd 3$; and the third in all cases where m^2z^2 is greater than 1 and less than $\frac{1}{2}n$, if n is an even number and very large. If n is not large this last rule cannot be much wanted, because, m decreasing continually as n is diminished, the value of z may in this case be taken large (and therefore a considerable interval had between $p/n - z$ and $p/n + z$), and yet the operation be carried on by the second rule; or mz not exceed $\surd 3$.

But in order to shew distinctly and fully the nature of the present problem, and how far Mr Bayes has carried the solution of it; I shall give the result of this solution in a few cases, beginning with the lowest and most simple.

Let us then first suppose, of such an event as that called M in the essay, or an event about the probability of which, antecedently to trials, we know nothing, that it has happened *once*, and that is is enquired what conclusion we may draw from hence with respect to the probability of it's happening on a *second* trial.

The answer is that there would be an odds of three to one for somewhat more than an even chance that it would happen on a second trial.

For in this case, and in all others where q is nothing, the expression

$$(n+1)\left\{\frac{X^{p+1}}{p+1} - \frac{x^{p+1}}{p+1}\right\} \qquad \text{or} \qquad X^{p+1} - x^{p+1}$$

gives the solution, as will appear from considering the first rule. Put therefore in this expression $p + 1 = 2$, $X = 1$ and $x = \frac{1}{2}$ and it will be $1 - (\frac{1}{2})^2$ or $\frac{3}{4}$; which shews the chance there is that the probability of an event that has happened once lies somewhere between 1 and $\frac{1}{2}$; or (which is the same) the odds that it is somewhat more than an even chance that it will happen on a second trial.*

In the same manner it will appear that if the event has happened twice, the odds now mentioned will be seven to one; if thrice, fifteen to one; and in general, if the event has happened p times, there will be an odds of $2^{p+1} - 1$ to one, for *more* than an equal chance that it will happen on further trials.

Again, suppose all I know of an event to be that it has happened ten times without failing, and the enquiry to be what reason we shall have to think we are right if we guess that the probability of it's happening in a single trial lies somewhere between $\frac{16}{17}$ and $\frac{2}{3}$, or that the ratio of the causes

*There can, I suppose, be no reason for observing that on this subject unity is always made to stand for certainty, and $\frac{1}{2}$ for an even chance.

of it's happening to those of it's failure is some ratio between that of sixteen to one and two to one.

Here $p + 1 = 11$, $X = \frac{16}{17}$ and $x = \frac{2}{3}$ and $X^{p+1} - x^{p+1} = (\frac{16}{17})^{11} - (\frac{2}{3})^{11} = 0.5013$ etc. The answer therefore is, that we shall have very nearly an equal chance for being right,

In this manner we may determine in any case what conclusion we ought to draw from a given number of experiments which are unopposed by contrary experiments. Everyone sees in general that there is reason to expect an event with more or less confidence according to the greater or less number of times in which, under given circumstances, it has happened without failing; but we here see exactly what this reason is, on what principles it is founded, and how we ought to regulate our expectations.

But it will be proper to dwell longer on this head.

Suppose a solid or die of whose number of sides and constitution we know nothing; and that we are to judge of these from experiments made in throwing it.

In this case, it should be observed, that it would be in the highest degree improbable that the solid should, in the first trial, turn any one side which could be assigned beforehand; because it would be known that some side it must turn, and that there was an infinity of other sides, or sides otherwise marked, which it was equally likely that it should turn. The first throw only shews that *it has* the side then thrown, without giving any reason to think that it has it any one number of times rather than any other. It will appear, therefore, that *after* the first throw and not before, we should be in the circumstances required by the conditions of the present problem, and that the whole effect of this throw would be to bring us into these circumstances. That is: the turning the side first thrown in any subsequent single trial would be an event about the probability or improbability of which we could form no judgment, and of which we should know no more than that it lay somewhere between nothing and certainty. With the second trial then our calculations must begin; and if in that trial the supposed solid turns again the same side, there will arise the probability of three to one that it has more of that sort of sides than of *all* others; or (which comes to the same) that there is somewhat in its constitution disposing it to turn that side oftenest: And this probability will increase, in the manner already explained, with the number of times in which that side has been thrown without failing. It should not, however, be imagined that any number of such experiments can give sufficient reason for thinking that it would *never* turn any other side. For, suppose it has turned the same side in every trial a million of times. In these circumstances there would be an improbability that it has *less* than 1,400,000 more of these sides than all others; but there would also be an improbability that it had *above* 1,600,000 times more. The chance for the latter is expressed by 1,600,000/1,600,001 raised to the millioneth power subtracted from unity, which is equal to 0.4647 etc and the chance for the former is equal to 1,400,000/1,400,001 raised to the same power, or to 0.4895; which, being both less than an equal chance, proves what I have said. But though it would be thus improbable that it had *above* 1,600,000 times more or *less* than 1,400,000 times *more* of these sides than of all others, it by no means follows that we have any reason for judging that the true proportion in this case lies somewhere between that of 1,600,000 to one and 1,400,000 to one. For he that will take the pains to make the calculation will find that there is nearly the probability expressed by 0.527, or but little more than an equal chance, that it lies somewhere between that of 600,000 to one and three millions to one. It may deserve to be added, that it is more probable that this proportion lies somewhere between that of 900,000 to 1 and 1,900,000 to 1 than between any other two proportions whose antecedents are to one another as 900,000 to 1,900,000, and consequents unity.

I have made these observations chiefly because they are all strictly applicable to the events and appearances of nature. Antecedently to all experience, it would be improbable as infinite to one, that any particular event, beforehand imagined, should follow the application of any one natural object to another; because there would be an equal chance for any one of an infinity of other events. But if we had once seen any particular effects, as the burning of wood on putting it into fire, or the falling of a stone on detaching it from all contiguous objects, then the conclusions to be drawn from any number of subsequent events of the same kind would be to be determined in the same manner with the conclusions just mentioned relating to the constitution of the solid I have supposed. In other words. The first experiment supposed to be ever made on any natural object would only inform us of one event that may follow a particular change in the circumstances of those objects; but it would not suggest to us any ideas of uniformity in nature, or give us the least reason to apprehend that it was, in that instance or in any other, regular rather than irregular in its operations. But if the same event has followed without interruption in any one or more subsequent experiments, then some degree of uniformity will be observed; reason will be given to expect the same success in further experiments, and the calculations directed by the solution of this problem may be made.

One example here it will not be amiss to give.

Let us imagine to ourselves the case of a person just brought forth into this world, and left to collect from his observation of the order and course of events what powers and causes take place in it. The Sun would, probably, be the first object that would engage his attention; but after losing it the first night he would be entirely ignorant whether he should ever see it again. He would therefore be in the condition of a person making a first experiment about an event entirely unknown to him. But let him see a second appearance or one *return* of the Sun, and an expectation would be raised in him of a second return, and he might know that there was an odds of 3 to 1 for *some* probability of this. This odds would increase, as before represented, with the number of returns to which he was witness. But no finite number of returns would be sufficient to produce absolute or physical certainty. For let it be supposed that he has seen it return at regular and stated intervals a million of times. The conclusions this would warrant would be such as follow. There would be the odds of the millioneth power of 2, to one, that it was likely that it would return again at the end of the usual interval. There would be the probability expressed by 0.5352, that the odds for this was not *greater* than 1,600,000 to 1; and the probability expressed by 0.5105, that it was not less than 1,400,000 to 1.

It should be carefully remembered that these deductions suppose a previous total ignorance of nature. After having observed for some time the course of events it would be found that the operations of nature are in general regular, and that the powers and laws which prevail in it are stable and permanent. The consideration of this will cause one or a few experiments often to produce a much stronger expectation of success in further experiments than would otherwise have been reasonable; just as the frequent observation that things of a sort are disposed together in any place would lead us to conclude, upon discovering there any object of a particular sort, that there are laid up with it many others of the same sort. It is obvious that this, so far from contradicting the foregoing deductions, is only one particular case to which they are to be applied.

What has been said seems sufficient to shew us what conclusions to draw from *uniform* experience. It demonstrates, particularly, that instead of proving that events will *always* happen agreeably to it, there will be always reason against this conclusion. In other words, where the course of nature has been the most constant, we can have only reason to reckon upon a recurrency of events proportioned to the degree of this constancy; but we can have no reason for thinking that there are no causes in nature which will *ever* interfere with the operations of the causes from which this constancy is derived, or no circumstances of the world in which it will fail. And if this is true, supposing our only *data* derived from experience, we shall find additional reason for thinking thus if we apply other principles, or have recourse to such considerations as reason, independently of experience, can suggest.

But I have gone further than I intended here; and it is time to turn our thoughts to another branch of this subject; I mean, to cases where an experiment has sometimes succeeded and sometimes failed.

Here, again, in order to be as plain and explicit as possible, it will be proper to put the following case, which is the easiest and simplest I can think of.

Let us then imagine a person present at the drawing of a lottery, who knows nothing of its scheme or of the proportion of *Blanks* to *Prizes* in it. Let it further be supposed, that he is obliged to infer this from the number of *blanks* he hears drawn compared with the number of *prizes*; and that it is enquired what conclusions in these circumstances he may reasonably make.

Let him first hear *ten* blanks drawn and *one* prize, and let it be enquired what chance he will have for being right if he guesses that the proportion of *blanks* to *prize* in the lottery lies somewhere between the proportions of 9 to 1 and 11 to 1.

Here taking $X = \frac{11}{12}$, $x = \frac{9}{10}$, $p = 10$, $q = 1$, $n = 11$, $E = 11$, the required chance, according to first rule, is $(n+1)E$ multiplied by the difference between

$$\left\{\frac{X^{p+1}}{p+1} - \frac{qX^{p+2}}{p+2}\right\} \quad \text{and} \quad \left\{\frac{x^{p+1}}{p+1} - \frac{qx^{p+2}}{p+2}\right\} =$$

$$12.11 \cdot \left\{\left[\frac{\left(\frac{11}{12}\right)^{11}}{11} - \frac{\left(\frac{11}{12}\right)^{12}}{12}\right] - \left[\frac{\left(\frac{9}{10}\right)^{11}}{11} - \frac{\left(\frac{9}{10}\right)^{12}}{12}\right]\right\} = 0 \cdot 07699 \text{ etc.}$$

There would therefore be an odds of about 923 to 76, or nearly 12 to 1 *against* his being right. Had he guessed only in general that there were less than 9 blanks to a prize, there would have been a probability of his being right equal to 0.6589, or the odds of 65 to 34.

Again, suppose that he has heard 20 *blanks* drawn and 2 *prizes*; what chance will he have for being right if he makes the same guess?

Here X and x being the same, we have $n = 22$, $p = 20$, $q = 2$, $E = 231$, and the required chance equal to

$$(n+1)E\left\{\left[\frac{X^{p+1}}{p+1} - \frac{qX^{p+2}}{p+2} + \frac{q(q-1)X^{p+3}}{2(p+3)}\right]\right.$$

$$\left.-\left[\frac{x^{p+1}}{p+1} - \frac{qx^{p+2}}{p+2} + \frac{q(q-1)x^{p+3}}{2(p+3)}\right]\right\} = 0.10843 \text{ etc.}$$

He will, therefore, have a better chance for being right than in the former instance, the odds against him now being 892 to 108 or about 9 to 1. But should he only guess in general, as before, that there were less than 9 blanks to a prize, his chance for being right will be worse; for instead of 0.6589 or an odds of near two to one, it will be 0.584, or an odds of 584 to 415.

Suppose, further, that he has heard 40 *blanks* drawn and 4 *prizes*; what will the before-mentioned chances be?

The answer here is 0.1525, for the former of these chances; and 0.527, for the latter. There will, therefore, now be an odds of only $5\frac{1}{2}$ to 1 against the proportion of blanks to prizes lying between 9 to 1 and 11 to 1; and but little more than an equal chance that it is less than 9 to 1.

Once more. Suppose he has heard 100 *blanks* drawn and 10 *prizes*.

The answer here may still be found by the first rule; and the chance for a proportion of blanks to prizes *less* than 9 to 1 will be 0.44109, and for a proportion *greater* than 11 to 1, 0.3082. It would therefore be likely that there were not *fewer* than 9 or *more* than 11 blanks to a prize. But at the same time it will remain unlikely* that the true proportion should lie between 9 to 1 and 11 to 1, the chance for this being 0.2506 etc. There will therefore be still an odds of near 3 to 1 against this.

From these calculations it appears that, in the circumstances I have supposed, the chance for being right in guessing the proportion of *blanks* to *prizes* to be nearly the same with that of the number of *blanks* drawn in a given time to the number of prizes drawn, is continually increasing as these numbers increase; and that therefore, when they are considerably large, this conclusion may be looked upon as morally certain. By parity of reason, it follows universally, with respect to every event about which a great number of experiments has been made, that the causes of its happening bear the same proportion to the causes of its failing, with the number of happenings to the number of failures; and that, if an event whose causes are supposed to be known, happens oftener or seldomer than is agreeable to this conclusion, there will be reason to believe that there are some unknown causes which disturb the operations of the known ones. With respect, therefore, particularly to the course of events in nature, it appears, that there is demonstrative evidence to prove that they are derived from permanent causes, or laws originally established in the constitution of nature in order to produce that order of events which we observe, and not from any of the powers of chance.† This is just as evident as it would be, in the case I

*I suppose no attentive person will find any difficulty in this. It is only saying that, supposing the interval between nothing and certainty divided into a hundred equal chances, there will be 44 of them for a less proportion of blanks to prizes than 9 to 1, 31 for a greater than 11 to 1, and 25 for some proportion between 9 to 1 and 11 to 1; in which it is obvious that, though one of these suppositions must be true, yet, having each of them more chances against them than for them, they are all separately unlikely.

†See Mr De Moivre's *Doctrine of Chances*, page 250.

have insisted on, that the reason of drawing 10 times more *blanks* than *prizes* in millions of trials, was, that there were in the wheel about so many more *blanks* than *prizes*.

But to proceed a little further in the demonstration of this point.

We have seen that supposing a person, ignorant of the whole scheme of a lottery, should be led to conjecture, from hearing 100 *blanks* and 10 prizes drawn, that the proportion of *blanks* to *prizes* in the lottery was somewhere between 9 to 1 and 11 to 1, the chance for his being right would be 0.2506 etc. Let [us] now enquire what this chance would be in some higher cases.

Let it be supposed that *blanks* have been drawn 1000 times, and prizes 100 times in 1100 trials.

In this case the powers of X and x rise so high, and the number of terms in the two series

$$\frac{X^{p+1}}{p+1} - \frac{qX^{p+2}}{p+2}\text{etc.} \qquad \text{and} \qquad \frac{x^{p+1}}{p+1} - \frac{qx^{p+2}}{p+2}\text{etc.}$$

become so numerous that it would require immense labour to obtain the answer by the first rule. 'Tis necessary, therefore, to have recourse to the second rule. But in order to make use of it, the interval between X and x must be a little altered. $\frac{10}{11} - \frac{9}{10}$ is $\frac{1}{110}$, and therefore the interval between $\frac{10}{11} - \frac{1}{110}$ and $\frac{10}{11} + \frac{1}{110}$ will be nearly the same with the interval between $\frac{9}{10}$ and $\frac{11}{12}$, only somewhat larger. If then we make the question to be; what chance there would be (supposing no more known than that blanks have been drawn 1000 times and prizes 100 times in 1100 trials) that the probability of drawing a blank in a single trial would lie somewhere between $\frac{10}{11} - \frac{1}{110}$ and $\frac{10}{11} + \frac{1}{110}$ we shall have a question of the same kind with the preceding questions, and deviate but little from the limits assigned in them.

The answer, according to the second rule, is that this chance is greater than

$$\frac{2\Sigma}{1 + 2Ea^pb^q + \dfrac{2Ea^pb^q}{n}}$$

and less than

$$\frac{2\Sigma}{1 - 2Ea^pb^q - 2E\dfrac{a^pb^q}{n}}$$

Σ being

$$\frac{(n+1)\surd(2pq)}{n\surd n}Ea^{p}b^{q}\left\{mz - \frac{m^3z^3}{3} + \frac{(n-2)m^5z^5}{2n.5} - \text{etc.}\right\}.$$

By making here $1000 = p$, $100 = q$, $1100 = n$, $\frac{1}{110} = z$,

$$mz = z\surd\left(\frac{n^3}{pq}\right) = 1.048808, \quad Ea^{p}b^{q} = \tfrac{1}{2}h\frac{\surd n}{\surd(Kpq)}^{r}$$

h being the ratio whose hyperbolic logarithm is

$$\frac{1}{12}\left[\frac{1}{n} - \frac{1}{p} - \frac{1}{q}\right] - \frac{1}{360}\left[\frac{1}{n^3} - \frac{1}{p^3} - \frac{1}{q^3}\right]$$
$$+ \frac{1}{1260}\left[\frac{1}{n^5} - \frac{1}{p^5} - \frac{1}{q^5}\right] - \text{etc.}$$

and K the ratio of the quadrantal arc to radius; the former of these expressions will be found to be 0.7953, and the latter 0.9405 etc. The chance enquired after, therefore, is greater than 0.7953, and less than 0.9405. That is; there will be an odds for being right in guessing that the proportion of blanks to prizes lies *nearly* between 9 to 1 and 11 to 1, (or *exactly* between 9 to 1 and 1111 to 99), which is greater than 4 to 1, and less than 16 to 1.

Suppose, again, that no more is known than that *blanks* have been drawn 10,000 times and *prizes* 1000 times in 11,000 trials; what will the chance now mentioned be!

Here the second as well as the first rule becomes useless, the value of mz being so great as to render it scarcely possible to calculate directly the series

$$\left\{mz - \frac{m^3z^3}{3} + \frac{(n-2)m^5z^5}{2n\cdot 5} - \text{etc.}\right\}$$

The third rule, therefore, must be used; and the information it gives us is, that the required chance is greater than 0.97421, or more than an odds of 40 to 1.

By calculations similar to these may be determined universally, what expectations are warranted by any experiments, according to the different number of times in which they have succeeded and failed; or what should be thought of the probability that any particular cause in nature, with which we have any acquaintance, will or will not, in any single trial, produce an effect that has been conjoined with it.

Most persons, probably, might expect that the chances in the specimen I have given would have been greater than I have found them. But this only shews how liable we are to error when we judge on this subject independently of calculation. One thing, however, should be remembered here; and that is, the narrowness of the interval between $\frac{9}{10}$ and $\frac{11}{12}$, or between $\frac{10}{11} + \frac{1}{110}$ and $\frac{10}{11} - \frac{1}{110}$. Had this interval been taken a little larger, there would have been a considerable difference in the results of the calculations. Thus had it been taken double, or $z = \frac{1}{55}$, it would have been found in the fourth instance that instead of odds against there were odds for being right in judging that the probability of drawing a blank in a single trial lies between $\frac{10}{11} + \frac{1}{55}$ and $\frac{10}{11} - \frac{1}{55}$.

The foregoing calculations further shew us the uses and defects of the rules laid down in the essay. 'Tis evident that the two last rules do not give us the required chances within such narrow limits as could be wished. But here again it should be considered, that these limits become narrower and narrower as q is taken larger in respect of p; and when p and q are equal, the exact solution is given in all cases by the second rule. These two rules therefore afford a direction to our judgment that may be of considerable use till some person shall discover a better approximation to the value of the two series in the first rule.*

But what most of all recommends the solution in this *Essay* is, that it is compleat in those cases where information is most wanted, and where Mr De Moivre's solution of the inverse problem can give little or no direction; I mean, in all cases where either p or q are of no considerable magnitude. In other cases, or when both p and q are very considerable, it is not difficult to perceive the truth of what has been here demonstrated, or that there is reason to believe in general that the chances for the happening of an event are to the chances for its failure in the same *ratio* with that of p to q. But we shall be greatly deceived if we judge in this manner when either p or q are small. And tho' in such cases the *Data* are not sufficient to discover the exact probability of an event, yet it is very agreeable to be able to find the limits between which it is reasonable to think it must lie, and also to be able to determine the precise degree of assent which is due to any conclusions or assertions relating to them.

*Since this was written I have found out a method of considerably improving the approximation in the second and third rules by demonstrating that the expression $2\Sigma/\{1 + 2Ea^p b^q + 2Ea^p b^q/n\}$ comes almost as near to the true value wanted as there is reason to desire, only always somewhat less. It seems necessary to hint this here; though the proof of it cannot be given.

Bibliography

Abraham, R. H., and Shaw, C. D. (1983). *Dynamics: The Geometry of Behavior*, Vol. 0: Manifolds and Mappings, Vol. 1: Periodic Behavior, Vol. 2: Chaotic Behavior, Vol. 3: Global Behavior, Vol. 4: Bifurcation Behavior. Santa Cruz, CA: Aerial Press.

Allais, M. (1953). "Le Comportement de L'Homme Rationnel Devant Le Risque: Critique Des Postulats et Axiomes de L'Ecole Américaine," *Econometrica*, **21**, 503–46. Translated in *Expected Utility Hypotheses and the Allais Paradox*, M. Allais and O. Hagen, eds. Dordrecht: Reidel, 1979.

Anscombe, F. J., and Aumann, R. J. (1963). "A Definition of Subjective Probability," *Ann. Math. Stat.*, **34**, 199–205.

Aykac, A., and Brumat, C., eds. (1976). *New Developments in the Applications of Bayesian Methods*. Amsterdam: North-Holland.

Barnett, V. (1982). *Comparative Statistical Inference*, 2nd edition. New York: Wiley.

Basu, D. (1964). "Recovery of Ancillary Information," *Sankhyā (A)*, **26**, 3–16.

Basu, D. (1988). *Statistical Information and Likelihood: A Collection of Critical Essays*, Lecture Notes in Statistics, **45**, New York: Springer-Verlag.

Bayes, T. (1763). "An Essay Towards Solving a Problem in the Doctrine of Chances," *Philos. Trans. R. Soc London*, **53**, 370–418. Reprinted in *Biometrika*, 1958, **45**, 293–315. It is reprinted, for the convenience of the reader, as Appendix 4 to this book.

Berge, P., Pomeau, Y., and C. Vidal (1987). *Order Within Chaos*, New York: John Wiley & Sons.

Berger, J. O. (1980). *Statistical Decision Theory*. New York: Springer-Verlag.

Berger, J. O. (1985). *Statistical Decision Theory and Bayesian Analysis*. New York: Springer-Verlag.

Berger, J. O., and Sellke, T. (1987). "Testing a Point Null Hypothesis: The Irreconcilability of *p*-Values and Evidence," *J. Am. Stat. Assoc.*, **82**, 397, 112–122.

Berger, J. O., and Wolpert, R. L. (1985). *The Likelihood Principle*, IMS Monograph Series, Vol. 6. Hayward: Calif.: Institute of Mathematical Statistics.

Bernardo, J. M. (1979). "Reference Posterior Distributions for Bayesian Inference" (with discussion), *J. R. Stat. Soc. (B)*, **41**(2), 113–147.

Bernardo, J. M. (1980). "A Bayesian Analysis of Classical Hypothesis Testing," in *Bayesian Statistics*, J. M. Bernardo, M. H. De Groot, D. V. Lindley, and A. F. M. Smith, eds. Valencia, Spain: University Press, pp. 605–618.

Bernardo, J. M., De Groot, M. H., Lindley, D. V., and Smith, A. F. M. (eds.) (1980). *Bayesian Statistics: Proceedings of the First International Meeting Held in Valencia (Spain)*. Valencia, Spain: University Press.

Bernardo, J. M., De Groot, M. H., Lindley, D. V., and Smith, A. F. M. (eds.) (1985). *Bayesian Statistics 2*. Amsterdam: North-Holland.

Berry, D. A., and Geisser, S. (1986). "Inference in Cases of Disputed Paternity," in *Statistics and the Law*, M. H. De Groot, S. E. Fienberg, J. B. Kadane, eds. New York: Wiley, pp. 353–382.

Birnbaum, A. (1962). "On the Foundations of Statistical Inference," *J. Am. Stat. Assoc.*, **57**, 269–306.

Blackwell, D. and Girshick, M. A. (1954). *Theory of Games and Statistical Decisions*, New York: John Wiley & Sons, Inc.

Boos, D. D., and Monahan, J. F. (1983). "Posterior Distributions from Bootstrapped Likelihoods," Department of Statistics, North Carolina State University, Raleigh, NC (manuscript, Dec. 1983).

Box, G. E. P., and Tiao, G. C. (1965). "Multiparameter Problems from a Bayesian Point of View," *Ann. Math. Stat.*, **36**, 1468–1482.

Box, G. E. P., and Tiao, G. C. (1973). *Bayesian Inference in Statistical Analysis*. Reading, MA: Addison-Wesley.

Broemeling, L. D. (1985). *Bayesian Analysis of Linear Models*. New York: Marcel Dekker.

Cacoullos, T. (1966). "Estimation of a Multivariate Density," *Ann. Inst. Stat. Math.* **2**(2), 179–189.

Carnap, R. and Jeffrey, R. C., eds. (1971). *Studies in Inductive Logic and Probability*, Vol. 1, Berkeley: University of California Press.

Casella, G., and Berger, R. L. (1987). "Reconciling Bayesian and Frequentist Evidence in the One-Sided Testing Problem," *J. Am. Stat. Assoc.*, **82**, 397, 106–111.

Cohen, A. M., Cutts, J. F., Fielder, R., Jones, D. E., Ribbans, J., and Stuart, E. (1973). *Numerical Analysis*. New York: Halsted Press, a division of John Wiley & Sons.

Davis, P. J., and Rabinowitz, P. (1967). *Numerical Integration*, Waltham, MA: Blaisdell.

de Finetti, B. (1937). "Le Prévision: ses lois logiques, ses sources subjectives," *Ann. Inst. Poincaré*, tome VII, fasc. 1, 1–68. Reprinted in *Studies in Subjective Probability*. Melbourne, FL: Krieger, 1980 (English translation).

de Finetti, B. (1974). *Theory of Probability* (Vols. 1 and 2). New York: Wiley.

De Groot, M. (1970). *Optimal Statistical Decisions*. New York: McGraw-Hill.

Dempster, A. P. (1967). "Upper and Lower Probabilities Induced by a Multivalued Mapping," *Ann. Math. Stat.*, **38**, 325–339.

Dempster, A. P. (1980). "Bayesian Inference in Applied Statistics," in *Bayesian Statistics*, J. M. Bernardo, M. H. De Groot, D. V. Lindley, and A. F. M. Smith, eds. Valencia, Spain: University Press, pp. 266–291.

Diaconis, P. (1977). "Finite Forms of de Finetti's Theorem on Exchangeability," *Synthese*, **36**, 271–281.

Diaconis, P. (1980). "Finite Exchangeable Sequences," *Ann. Prob.*, **8**(4), 745–764.

Diaconis, P., and Freedman, D. (1980). "de Finetti's Generalizations of Exchangeability," in *Studies in Inductive Logic and Probability*, R.C. Jeffrey, ed. Berkeley: University of California Press, 223–249.

Diaconis, P., and Freedman, D. (1984). "Partial Exchangeability and Sufficiency," in *Statistics: Applications and New Directions*, J. K. Ghosh and J. Roy, eds. Calcutta; Indian Statistical Institute, pp. 205–236.

Dickey, J. M. (1973). "Scientific Reporting and Personal Problems: Student's Hypothesis," *J. R. Stat. Soc. (B)*, **35**, 285–305.

Dickey, J. M., Lindley, D. V., and Press, S. J. (1985). "Bayesian Estimation of the Dispersion Matrix of a Multivariate Normal Distribution," *Communications in Statistics: Theory & Methods*, **14**(5). Special section on Bayesian statistics, 1019–1034.

Dunford, N., and Schwartz, J. T. (1965). *Linear Operators*, Part I. New York: Wiley.

Edwards, W., Lindman, H., and Savage, L. J. (1963). "Bayesian Statistical Inference for Psychological Research," *Psych. Rev.*, **70**, 193–242.

Efron, B. (1979). "Bootstrap Methods: Another Look at the Jackknife," *Ann. Stat.*, **7**, 1–26.

Efron, B. (1982). *The Jackknife, the Bootstrap and Other Resampling Plans*, CBMS-NSF Regional Conference Series in Applied Mathematics. Philadelphia: Society for Industrial and Applied Mathematics.

Efron, B. (1986). "Why Isn't Everyone a Bayesian?" *Am. Stat.* **40**(1), 1–11.

Efron, B., and Morris, C. (1973). "Stein's Estimation Rule and Its Competitors—An Empirical Bayes Approach," *J. Am. Stat. Assoc.*, **68**, 117–130.

Efron, B. and Morris, C. (1977). "Stein's Paradox in Statistics," *Scientific American Magazine*, May, 1977.

Erdelyi, A. (1956). *Asymptotic Expansions*. New York: Dover Publications.

Ferguson, T. S. (1967). *Mathematical Statistics*. New York: Academic Press.

Fienberg, S. E., and Zellner, A., eds. (1975). *Studies in Bayesian Econometrics and Statistics*. Amsterdam: North-Holland.

Fishburn, P. C. (1981). "Subjective Expected Utility: A Review of Normative Theories," *Theory & Decision*, **13**, 139–199.

Fishburn, P. C. (1986). "The Axioms of Subjective Probability," *Statistical Science*, Vol. 1, No. 3, 335–358.

Fraser, D. A. S. (1968). *The Structure of Inference*. New York: Wiley.

Freudenburg, W. R. (1988). "Perceived Risk, Real Risk: Social Science and the Art of Probabilistic Risk Assessment," *Science*, Vol. 242, 70 ct., 1988, 44–49.

Geisser, S. (1964). "Posterior Odds for Multivariate Normal Classifications," *J. R. Stat. Soc.* (*B*), **26**, 69–76.

Geisser, S. (1966). "Predictive Discrimination," in *Multivariate Analysis*, P. R. Krishnaiah, ed. New York: Academic Press, pp. 149–163.

Geisser, S. (1967). "Estimation Associated with Linear Discriminants," *Ann. Math. Stat.*, **38**, 807–817.

Geisser, S. (1980). "A Predictive Primer," in *Bayesian Analysis in Econometrics and Statistics*, A. Zellner, ed. Amsterdam: North-Holland.

Geisser, S., and Cornfield, J. (1963). "Posterior Distributions for Multivariate Normal Parameters," *J. R. Stat. Soc* (*B.*), **25**, 368–376.

Gleick, James (1987). *Chaos: Making a New Science*, New York: Viking Penguin, Inc.

Goel, P. K. (1987). "Software for Bayesian Analysis: Current Status and Additional Needs," Technical Report No. 366, Department of Statistics, The Ohio State University, Columbus, OH.

Good, I. J. (1950). *Probability and the Weighting of Evidence*. London: Griffin Publishing Co.

Good, I. J. (1965). *The Estimation of Probabilities: An Essay on Modern Bayesian Methods*, Cambridge, Mass.: MIT Press.

Good, I. J. (1983). *Good Thinking, The Foundations of Probability and Its Applications*, Minneapolis: The University of Minnesota Press.

Grayson, C. J. (1960). *Decisions Under Uncertainty: Drilling Decisions by Oil and Gas Operators*, Boston, Mass: Harvard Business School, Div. of Research.

Haavelmo, T. (1947). "Methods of Measuring the Marginal Propensity to Consume," *J. Am. Stat. Assoc.*, **42**, p. 88.

Hartigan, J. (1964). "Invariant Prior Distributions," *Ann. Math. Stat.*, **35**, 836–845.

Hartigan, J. (1983). *Bayes Theory*. New York: Springer-Verlag.

Heisenberg, W. (1927). "Über den anschaulichen Inhalt der quantentheoretischen Kinematik und Mechanik," *Zietschrift für Physik*, Vol. 43, No. 2, 172–198.

Hewitt, E., and Savage, L. (1955). "Symmetric Measures on Cartesian Products," *Trans. Am. Math. Soc.*, **80**, 470–501.

Hogarth, R. (1980). *Judgement and Choice*, New York: John Wiley & Sons, Inc.

Huber, P. J. (1977). *Robust Statistical Procedures*, CBMS Regional Conference Series in Applied Mathematics, No. 27. Philadelphia: Society for Industrial and Applied Mathematics.

Hyde, C. C., and Johnston, I. M. (1979). "On Asymptotic Posterior Normality for Stochastic Processes," *J. R. Stat. Soc.* (*B*), **41**, 184–189.

Iversen, G. R. (1984). *Bayesian Statistical Inference*. Beverly Hills, CA: Sage Publications, Inc.

James, W., and Stein, C. (1960). "Estimation with Quadratic Loss," *Fourth Berkeley Symposium on Mathematical Statistics and Probability*, Berkeley: University of California Press, pp. 361–379.

Jaynes, E. (1983). *Papers on Probability, Statistics, and Statistical Physics*, R. D. Rosenkrantz, ed. Dordrecht, Holland: Reidel.

Jeffrey, R. C., ed. (1980). *Studies in Inductive Logic and Probability*, Vol. 2, Berkeley: University of California Press.

Jeffreys, H. (1961). *Theory of Probability*, 3rd edition (1st edition, 1939, 2nd edition, 1948). Oxford: Clarendon Press.

Joos, G. (1934). *Theoretical Physics*, translated from the first German edition by I. M. Freeman. New York: Hafner.

Kadane, J. B. (ed.) (1984). *Robustness of Bayesian Analysis*. Amsterdam: North-Holland.

Kadane, J. B., Dickey, J. M., Winkler, R. L., Smith, W. S., and Peters, S. C. (1980). "Interactive Elicitation of Opinion for a Normal Linear Model," *J. Am. Stat. Assoc.*, **75**, 845–854.

Kahneman, D., Slovic, P., and Tversky, A. (1982). *Judgement Under Uncertainty: Heuristics and Biases*. Cambridge: Cambridge University Press.

Kanji, G. K., ed. (1983). *J. Inst. Stat.*, **32** (1 and 2); entire issue.

Kass, R. E. (May, 1985). "Exact Inferences About Principal Components and Related Quantities Using Posterior Distributions Calculated By Simulation," paper presented at 13*th* NSF/NBER Seminar on Bayesian Inference in Econometrics. (For information contact R. E. Kass, Dept. of Stat., Carnegie Mellon University.)

Keynes, J. M. (1921). *A Treatise on Probability*, London: MacMillan.

Klein, R., and Press, S. J. (1987), "Spatial Structure in Bayesian Classification," Tech. Rept. No. 157, Dept. of Stat., Univ. of Calif., Riverside, CA; and *Communications in Statistics*, 1989, in press.

Klein, R., and Press, S. J. (1988). "Bayesian Contextual Classification With Neighbors Correlated With Training Data," Tech. Rept. No. 167, Dept. of Stat., Univ. of Calif., Riverside, CA; and in *Bayesian and Likelihood Methods in Statistics & Econometrics, Essays in Honor of George A. Barnard*, eds. J. Hodges, S. Geisser, S. J. Press, and A. Zellner, Amsterdam: North Holland Pub. Co., 1989, in press.

Kloek, T., and Van Dijk, H. K. (1978). "Bayesian Estimates of Equation System Parameters: An Application of Integration by Monte Carlo," *Econometrica*, **46**, 1–19.

Koch, G., and Spizzichino, F. (1982). *Exchangeability in Probability and Statistics*. Amsterdam: North-Holland.

Kolmogoroff, A. N. (1933). Cited in Kolmogoroff, A. N., and Fowin, S. V. (1961), *Measure, Lebesgue Integrals, and Hilbert Space*. New York: Academic Press.

Kyburg, H. E., Jr., and Smokler, H. E., eds. (1980). *Studies in Subjective Probability*. New York: Krieger.

Laplace, P. S. (1774). "Mémoire sur la probabilité des causes par les évenemens," *Mem. Acad. R. Sci. Presentés par Divers Savans*, **6**, 621–656.

Laplace, P. S. (1774). "Memoir on the Probability of the Causes of Events," *Mémoires de Mathématique et de Physique, Presentés a l'Académie Royale des Sciences, par divers Savans et lûs dans ses Assemblés*, Tome Sixième, 621–656. Reprinted in *Laplace's Oeuvres Complètes*, **8**, 27–65. Translated by Stephen M. Stigler, and reprinted in translation in *Statistical Science*, 1986, Vol. 1, No. 3, 359–378.

Laplace, P. S. (1812). *Theorie Analytique des Probabilités*, Paris: Courcier. The second, third, and fourth editions appeared in 1814, 1818, and 1820, respectively. It is reprinted in *Oeuvres Completes de Laplace*, Vol. VII, 1847. Paris: Gauthier-Villars.

Laplace, P. S. (1814). *Essai Philosophique sur les Probabilités*, Paris. This book went through five editions (the fifth was in 1825) revised by Laplace. The sixth edition appeared in English translation by Dover Publications, New York, in 1951. While this philosophical essay appeared separately in 1814, it also appeared as a preface to his earlier work, *Theorie Analytique des Probabilités*.

Leamer, E. E. (1978). *Specification Searches*. New York: Wiley.

Le Cam, L. (1956). "On the Asymptotic Theory of Estimation and Testing Hypotheses," *Proceedings of the Third Berkeley Symposium on Mathematical Statistics and Probability* Vol. 1, Berkeley: University of California Press, pp. 129–156.

Lewis, R. (1988). "DNA Fingerprints: Witness for the Prosecution," *Discover Magazine*, June, 1988, Vol. 9, No. 6, 44–53.

Lindley, D. V. (1957). "A Statistical Paradox," *Biometrika*, **44**, 187–192.

Lindley, D. V. (1965). *Introduction to Probability and Statistics* (Part 1—Probability and Part 2—Inference). Cambridge: Cambridge University Press.

Lindley, D. V. (1972). *Bayesian Statistics: A Review*. Philadelphia: Society for Industrial and Applied Mathematics.

Lindley, D. V. (1976). "Bayesian Statistics," in *Foundations of Probability Theory, Statistical Inference, and Statistical Theories of Science*, Vol. II, W. L. Harper and C. A. Hooker, eds. Boston: Reidel, pp. 353–363.

Lindley, D. V. (1980). "Approximate Bayesian Methods," in *Bayesian Statistics*, J. M. Bernardo, M. H. De Groot, D. V. Lindley, and A. F. M. Smith, eds. Valencia, Spain: Valencia Press, pp. 223–245.

Lindley, D. V. (1985). *Making Decisions*. London: Wiley.

Lindley, D. V., and Smith, A. F. M. (1972). "Bayes Estimates for the Linear Model," *J. R. Stat. Soc. (B)*, **34**, 1–41.

Lindley, D. V., and Phillips L. D. (1976). "Inference for a Bernoulli Process (a Bayesian View)," *Am. Stat.*, **30**(3), 112–119.

Lindley, D. V., and Novick, M. R. (1982). "The Role of Exchangeability in Inference," *Ann. Stat.*, **9**(1), 45–58.

Lo, A. Y. (1984). "On a Class of Bayesian Nonparametric Estimates: I. Density Estimates," *Ann. Stat.*, **12**, 351–357.

Maistrov, L. E. (1974). *Probability Theory: A Historical Sketch*, translated by S. Kotz, New York: Academic Press.

Mandelbrot, B. B. (1977a). *Fractals, Form, Chance, and Dimension*, San Francisco: W. H. Freeman and Company.

Mandelbrot, B. B. (1977b). *The Fractal Geometry of Nature*, New York: W. H. Freeman and Company.

Maritz, J. S. (1970). *Empirical Bayes Methods*, London: Methuen & Co.

Martz, H. F., and Waller, R. A. (1982). *Bayesian Reliability Analysis*. New York: Wiley.

McKean, K. (1987). "The Orderly Pursuit of Pure Disorder," *Discover Magazine*, published by Time Inc., New York, Jan., 1987, 72–81.

Moon, F. C. (1987). *Chaotic Vibrations*, New York: John Wiley.

Mosteller, F., and Wallace, D. L. (1964). *Inference and Disputed Authorship: The Federalist*. Reading, MA: Addison-Wesley.

Mosteller, F., and Wallace, D. L. (1984). *Applied Bayesian and Classical Inference*, New York: Springer-Verlag.

Naylor, J. C. (1982). "Some Numerical Aspects of Bayesian Inference," Unpublished Ph.D. thesis, University of Nottingham.

Naylor, J. C., and Smith, A. F. M. (1982). "Applications of a Method for the Efficient Computation of Posterior Distributions," *App. Stat.*, **31**(3), 214–225.

Naylor, J. C., and Smith, A. F. M. (1983). "A Contamination Model in Clinical Chemistry," in *Practical Bayesian Statistics*, A. P. Dawid and A. F. M. Smith, eds. Harlow: England, Longman.

Pagels, H. R. (1982). *The Cosmic Code*. New York: Bantam Books.

Parzen, E. (1962). "On Estimation of a Probability Density Function and Mode," *Ann. Math. Stat.*, **33**, 1065–1076.

Pearson, E. (1978). *The History of Statistics in the 17th and 18th Centuries*. New York: Macmillan.

Peitgen, H.-O. and Richter, P. H. (1986). *The Beauty of Fractals, Images of Complex Dynamical Systems*, Heidelberg: Springer-Verlag.

Peitgen, H.-O. and Saupe, D. editors (1988). *The Science of Fractal Images*, New York: Springer-Verlag.

Phillips, L. D. (1974). *Bayesian Statistics for Social Scientists*. New York: Crowell.

Pollard, W. E. (1986). *Bayesian Statistics for Evaluation Research*, Beverly Hills: Sage Publications, Inc.

Popper, K. R. (1968). *The Logic of Scientific Discovery*. New York: Harper & Row.

Pratt, J. W. (1965). "Bayesian Interpretation of Standard Inference Statements," *J. R. Stat. Soc. (B)*, **27**, 169–203.

Pratt, J. W., Raiffa, H., and Schaifer, R. (1964). "The Foundations of Decision Under Uncertainty: An Elementary Exposition," *J. Am. Stat. Assoc.*, **59**, 353–375.

Pratt, J. W., Raiffa, H., and Schaifer, R. (1965). *Introduction to Statistical Decision Theory* (Preliminary Edition). New York: McGraw-Hill.

Press, S. J. (1978). "Qualitative Controlled Feedback for Forming Group Judgments and Making Decisions," *J. Am. Stat. Assoc.*, **73**, 363, 526–535.

Press, S. J. (1980a). "Bayesian Computer Programs," in *Bayesian Analysis in Econometrics and Statistics*, A. Zellner, ed. Amsterdam: North–Holland.

Press, S. J. (1980b). "Bayesian Inference in MANOVA," in *Handbook of Statistics*, P. R. Krishnaiah, ed. Amsterdam: North-Holland.

_____ (1980c). "Bayesian Inference in Group Judgment Formulation and Decision Making Using Qualitative Controlled Feedback," in *Bayesian Statistics*, J. M. Bernardo, M. H. De Groot, D. V. Lindley, A. F. M. Smith (eds.), 383–430.

_____ (1980d). "Multivariate Group Judgments by Qualitative Controlled Feedback," in *Multivariate Analysis V*, (P. R. Krishnaiah, ed.), New York: North Holland Pub. Co., 581–591.

Press, S. J. (1982). *Applied Multivariate Analysis: Using Bayesian and Frequentist Methods of Inference*. Melborne, FL: Krieger.

Press, S. J. (1983). "Group Assessment of Multivariate Prior Distributions," *Tech. Forecast. Soc. Change*, **23**, 247–259.

Press, S. J. (1985a). "Multivariate Group Assessment of Probabilities of Nuclear War," in *Bayesian Statistics 2*, J. M. Bernardo, M. H. De Groot, D. V. Lindley, and A. F. M. Smith, eds. Amsterdam: North-Holland, pp. 425–462.

Press, S. J. (1985b). "Multivariate Analysis (Bayesian)," in *Encyclopedia of Statistical Sciences*, S. Kotz and N. L. Johnson, eds., New York: John Wiley & Sons, Inc., Vol. 6, pp. 16–20.

Press and Klein (see Klein and Press).

Press, S. J., and Shigemasu, K. (1985). "Bayesian MANOVA and MANOCOVA Under Exchangeability," *Comm. Stat.*, special issue devoted to recent developments in Bayesian inference.

Press, S. J., and Shigemasu, K. (1989). "Bayesian Inference In Factor Analysis," Tech. Rept. No. 176, Dept. of Statistics, Univ. of Calif., Riverside; and in *Contributions to Probability and Statistics: Essays in Honor of Ingram Olkin*, L. Gleser, M. Perlman, S. J. Press, A. Sampson, eds. New York: Springer-Verlag, 1989, in press.

Press, S. J., and Yang, C. (1975). "A Bayesian Approach to Second Guessing Undecided Respondents," *J. Am. Stat. Assoc.*, **69**, 58–67.

_____, Ali, M. W., and Yang, E. (1979). "An Empirical Study of a New Method for Forming Group Judgments: Qualitative Controlled Feedback," *Technological Forecasting & Social Change*, **15**, 171–189.

Raiffa, H., and Schlaifer, R. (1961). *Applied Statistical Decision Theory*. Graduate School of Business Administration. Boston, MA: Harvard University.

Ramsey, F. P. (1926). "Truth and Probability," in *The Foundations of Mathematics and Other Logical Essays (1931)*, R. B. Braithwaite, ed. by permission of The Humanities Press, New York, and Routledge and Kegan Paul Ltd., London. Reprinted in Kyburg and Smokler (1980). *Studies in Subjective Probability*. Melbourne, FL: Krieger.

Rényi, A. (1970). *Probability Theory*. New York: American Elsevier.

Robbins, H. (1955). "An Empirical Bayes Approach to Statistics," *Proc. (Third) Berkeley Symp. Math. Stat. Probab.*, **1**, 157–163.

Roberts, H. (1978). "Bayesian Inference," in *International Encyclopedia of Statistics*, edited by W. H. Kruskal and J. M. Tanur, New York, The Free Press, a Division of Macmillan Pub. Co., Inc., pp. 9–16.

Robinson, G. K. (1975). "Some Counterexamples to the Theory of Confidence Intervals," *Biometrika*, **62**(1), 155–161.

Roethlisberger, F. W., and Dickson, W. J. (1939; republished in 1961). *Management and the Worker: An Account of a Research Program Conducted by the Western Electric Hawthorne Works, Chicago*. Boston: Harvard University Press (published by John Wiley & Sons in 1961 as a paperback).

Rosenkrantz, R. D. (1977). *Inference, Method and Decision: Towards A Bayesian Philosophy of Science*. Dordrecht, Holland: Reidel.

Rubin, D. B. (1981). "The Bayesian Bootstrap," *Ann. Stat.*, **9**, 130–134.

Rubin, D. B. (1987). *Multiple Imputation for Nonresponse in Surveys*. New York: Wiley.

Salzer, H. E., Zucker, R., and Capuano, R. (1952). "Tables of the Zeros and Weight Factors of the First Twenty Hermite Polynomials," *J. Res. Nat. Bur. Stand.*, **48**, 111–116.

Savage, L. J. (1954). *The Foundations of Statistics*. New York: Wiley.

Savage, L. J. et al. (1962). *The Foundations of Statistical Inference*. New York: Wiley (Methuen & Co. London.)

Savage, L. J. (1981). *The Writings of Leonard Jimmie Savage—A Memorial Collection*, Washington, D.C.: The American Statistical Association and the Institute of Mathematical Statistics.

Shafer, G. (1976). *A Mathematical Theory of Evidence*. Princeton: Princeton University Press.

Shafer, G. (1982). "Lindley's Paradox," *J. Am. Stat. Assoc.*, **77**(378), 325–351.

Silverman, B. W. (1986). *Density Estimation for Statistics and Data Analysis*, New York: Chapman and Hall.

Skyrms, B. (1984). *Pragmatics and Empiricism*. New Haven, CT: Yale University Press.

Skyrms, B. (1989). *Deliberational Dynamics*, forthcoming.

Smith, R. L., and Naylor, J. C. (1984). "A Comparison of Maximum Likelihood and Bayesian Estimators for the Three Parameter Weibull Distribution," Technical Report, Imperial College, London.

Smith, A. F. M., Skene, A. M., Shaw, J. E. H., Naylor, J. C., and Dransfield, M. (1985). "The Implementation of the Bayesian Paradigm," *Commun. Stat. Theor. Methods*, **14**(5), 1079–1102.

Stael von Holstein, C-A. S. (1970). *Assessment and Evaluation of Subjective Probability Distributions*. Stockholm: The Economic Research Institute.

Stein, C. (1965). "Approximation of Improper Prior Measures," in *Bernoulli, Bayes, Laplace*, J. Neyman and L. LeCam, eds. New York: Springer-Verlag.

Stewart, L. T. (1979). "Multiparameter Univariate Bayesian Analysis," *J. Am. Stat. Assoc.*, **74**, 684–693.

Stewart, L. T. (1983). "Bayesian Analysis Using Monte Carlo Integration—A Powerful Methodology for Handling Some Difficult Problems," in *Practical Bayesian Statistics*, A. P. Dawid, and A. F. M. Smith, eds. Harlow, England: Longman.

Stewart, L. T. (1984). "Multiparameter Bayesian Inference Using Monte Carlo Integration—Some Techniques for Bivariate Analysis," *Bayesian Statistics 2*, J. M. Bernardo, M. H. De Groot, D. V. Lindley, and A. F. M. Smith, eds. Amsterdam: North-Holland.

Stigler, S. M. (1982). "Thomas Bayes and Bayesian Inference," *J. R. Stat. Soc. (A)*, **145**(2), 250–258.

Stigler, S. M. (1983). "Who Discovered Bayes Theorem," *Am. Stat.*, **37**(4), 290–296.

Stigler, S. M. (1986). *The History of Statistics*. Cambridge, MA: The Belknap Press of Harvard University Press.

Taipia, R. A., and Thompson, J. R. (1982). *Nonparametric Probability Density Estimation*. Baltimore, MD: The Johns Hopkins University Press.

Thompson, J. M. T. and Stewart, H. B. (1986). *Nonlinear Dynamics and Chaos*, New York: John Wiley & Sons.

Tierney, L., and Kadane, J. B. (1984). "Accurate Approximations for Posterior Moments and Marginals," Technical Report No. 431, School of Statistics, University of Minnesota. Also see Tierney, L., and Kadane, J. B. (1986) *J. Am. Stat. Assoc.*, **81**, 82–86.

Tierney, L., Kass, R. E., and Kadane, J. B. (1988). "Fully Exponential Laplace Approximations of Expectations and Variances of Non-positive Functions," Technical Report No. 418, Department of Statistics, Carnegie-Mellon University.

Tversky, A., and Kahneman, D. (1974). "Judgment Under Uncertainty: Heuristics and Biases," *Science*, **185**, 1124–1131.

Van Dijk, H. K., and Kloek, T. (1980). "Further Experience in Bayesian Analysis Using Monte Carlo Integration," *J. Econometrics*, **14**, 307–328.

Van Dijk, H. K., and Kloek, T. (1983). "Monte Carlo Analysis of Skew Distributions: An Illustrative Econometric Example," in *Practical Bayesian Statistics*, A. P. Dawid and A. F. M. Smith, eds. Harlow, England: Longman.

Van Dijk, H. K., and Kloek, T. (1984). "Experiments with Some Alternatives For Simple Importance Sampling in Monte Carlo Integration," *Bayesian Statistics 2*, J. M. Bernardo, M. H. De Groot, D. V. Lindley, and A. F. M. Smith, eds. Amsterdam: North-Holland.

Varian, H. R. (1975), "A Bayesian Approach to Real Estate Assessment," in *Studies in Bayesian Econometrics and Statistics in Honor of Leonard J. Savage*, S. E. Fienberg, and A. Zellner, eds. Amsterdam: North Holland Pub. Co., 195–208.

Vijn, P. (1980). *Prior Information in Linear Models*. Rijksuniversiteit te Groningen, Groningen, The Netherlands.

Villegas, C. (1964). "On Qualitative Probability σ-Algebras," *Ann. Math. Stat.*, **35**, 1787–1796.

Villegas, C. (1969). "On the a priori Distribution of the Covariance Matrix," *Ann. Math. Statist.*, **40**, 1098–1099.

Villegas, C. (1977). "On the Representation of Ignorance," *J. Am. Stat. Assoc.*, **72**, 651–654.

von Mises, R. (1957). *Probability, Statistics, and Truth*, second revised English edition developed by Hilda Geiringer. New York: Macmillan.

von Mises, R. (1964). *The Mathematical Theory of Probability and Statistics*, developed by Hilda Geiringer. New York: Academic Press.

von Neumann, J., and Morgenstern, O. (1947). *Theory of Games and Economic Behavior*, 2nd edition. Princeton, NJ: Princeton University Press.

Wald, A. (1950). *Statistical Decision Functions*. New York: Wiley.

Winkler, R. L. (1972). *Introduction to Bayesian Inference and Decision*. New York: Holt, Rinehart and Winston.

Zellner, A. (1971). *An Introduction to Bayesian Inference in Econometrics*. New York: John Wiley & Sons.

Zellner, A. (ed.) (1980). *Bayesian Analysis in Econometrics and Statistics. Essays in Honor of Harold Jeffreys*. Amsterdam: North-Holland.

Zellner, A. (1985). "Bayesian Econometrics," *Econometrica*, **53**(2), 253–269.

Zellner, A. (1986). "On Assessing Prior Distributions and Bayesian Regression Analysis with *g*-Prior Distributions," *Bayesian Inference and Decision Techniques: Essays in Honor of Bruno de Finetti*, P. Goel and A. Zellner, eds. New York: North-Holland, pp. 233–243.

Zellner, A. (1986). "Bayesian Estimation and Prediction Using Asymmetric Loss Functions," *J. Am. Stat. Assoc.*, **81**, 446–451.

Zellner, A. (1987). "Bayesian Inference," in *The New Palgrave: A Dictionary of Economics*, Vol. 1, John Eatwell, Murray Milgate, and Peter Newman, eds. London: The MacMillan Press, Ltd., 208–218.

Author Index

Subject Index